Nanomaterials for Energy Applications

Topical Conference at the 2010 AIChE Annual Meeting

Salt Lake City, Utah, USA
7-12 November 2010

ISBN: 978-1-61782-138-7

Printed from e-media with permission by:

Curran Associates, Inc.
57 Morehouse Lane
Red Hook, NY 12571

Some format issues inherent in the e-media version may also appear in this print version.

Printed by Curran Associates, Inc. (2011)

AIChE
3 Park Avenue
New York, NY 10016-5991

Phone: (203) 702-7660
Fax: (203) 775-5177

www.aiche.org

Additional copies of this publication are available from:

Curran Associates, Inc.
57 Morehouse Lane
Red Hook, NY 12571 USA
Phone: 845-758-0400
Fax: 845-758-2634
Email: curran@proceedings.com
Web: www.proceedings.com

TABLE OF CONTENTS

Polymeric Composite Enhanced by Carbon Nanotube Yarns

Mei Zhang, Hang Zhang, Richard Liang, Chuck Zhang, and Ben Wang*
Department of Industrial and Manufacturing Engineering, High-Performance Materials Institute, FAMU-FSU College of Engineering, Tallahassee, FL (USA)
**Corresponding author: mzhang@eng.fsu.edu*

Abstract: The processes for fabricating polymer/carbon nanotube (CNT) composites were explored and developed. It is found that using CNT yarns as filler is an effective way to control the CNT distribution and content in composite. The reinforcement is significant. When using PMMA (Poly(methyl methacrylate)) as polymeric matrix, the modulus and strength of the composite increase linearly with the increase of CNT content. For the 17 wt% CNT-containing PMMA/CNT-yarn composite, compared to the neat PMMA sheet, the tensile strength and modulus of the composite have a 5-fold and 10-fold improvement, respectively. It is demonstrated that using the highly ordered CNT assemblies as filler is an efficient way to improve the properties of polymeric composite.

1. Introduction

Carbon Nanotubes (CNTs) belong to a class of nanomaterial that has remarkable physical and mechanical properties.[1] Their superlative mechanical properties make them the filler material for composite reinforcement. The improvement of the mechanical properties of nanocomposites including CNT have been reported.[2, 3] However, the mechanical properties of the composite made by dispersing CNTs in polymer matrix are much lower than expected. The reasons for the low enhancement are mainly due to the difficulties in uniform dispersion, alignment, and interconnections between CNTs and polymer matrix. When the polymer is poly(methyl methacrylate) (PMMA), several dispersion methods such as melt blending,[4] in-situ polymerization,[5] solution mixing (solvent evaporation),[6, 7] and mold injection[8] have been proposed. Those methods showed improvement in mechanical properties of the PMMA/CNT

1. M.S. Dresselhaus, G. Dresselhaus, P.Avouris, "Carbon nanotubes: synthesis, structure, properties and applications," *Springer Verlag*, the First Edition, 2001.
2. T. W. Chou, E. T. Thostenson, Z. F. Ren. "Advances in the science and technology of carbon nanotubes and their composites: A review," *Composites Science and Technology*, **61**, 1899 (2001).
3. J. N. Coleman, U. Khan, W. J. Blau, and Y. K. Gunko, "Small but strong: A review of the mechanical properties of carbon nanotube-polymer composites," *Carbon*, **44**, 1624 (2006).
4. Z. Jin, K. P. Pramoda, G. Xu, and S. H. Goh, "Dynamic Mechanical Behavior of Melt-Processed Multi-Walled Carbon Nanotube/Poly(Methyl Methacrylate) Composites," *Chemical Physics Letters*, **337**, 43-47 (2001).
5. Z. Jia, Z. Wang, C. Xu, J. Liang, B. Wei, D. Wu, S. Zhu, "Study on Poly(Methyl Methacrylate)/Carbon Nanotube Composites," *Materials Science and Engineering* : **A271**, 395-400 (1999).
6. C. Stéphan, T. P. Nguyen, M. Lamy de la Chapelle, S. Lefrant, C. Journet, and P. Bernier, "Characterization of Singlewalled Carbon Nanotubes-PMMA Composites," *Synthetic Metals*, **108**, 139-149 (2000).
7. K. H. Kim and W. H. Jo, "Improvement of Tensile Properties of Poly(Methyl Methacrylate) by Dispersing Multi-walled Carbon Nanotubes Functionalized with Poly(3-Hexylthiophene)-Graft-Poly(Methyl Methacrylate)," *Composites Science and Technology* **68**, 2120-2124 (2008).
8. W.-J. Lee, S.-E. Lee, and C.-G. Kim, "The mechanical properties of MWNT/PMMA nanocomposites fabricated by modified injection molding," *Composite Structures*, **76**, 406–410 (2006).

composites. However, the effects of CNT reinforcement were small. Instead of dispersing CNTs into the PMMA matrix, we used solid-state processes to assemble CNTs into highly ordered structures, CNT yarn and yarn array, and used them as filler to reinforce PMMA. This approach allows us to well control the alignment of CNTs and to have high CNT content.

2. Experimental

Materials used: The PMMA pellets (Evonik CYRO LLC ACRYLITE 8N Acrylic) were used to make PMMA sheet. The PMMA used has features of high strength, high heat resistance, and high clarity. The density and the melting point of the PMMA are 1.17 g/cm^3 and 220°C, respectively. The CNT yarns were twist spun from the CNT forest synthesized by chemical vapor deposition.[9] The twist spinning process assembles the CNTs into a denser and highly ordered structure without lowering their aspect ratio or adding chemical additives.

Fabrication of PMMA/CNT-yarn composite: The PMMA sheet was made by using PMMA pellets through hot-press process. The PMMA/CNT-yarn composites were made by applying CNT yarns on top of a PMMA sheet to form an array, and then set it into a mold and placed the mold in a hot press apparatus. The procedure is to raise the temperature without applying pressure for 30 min till the plates temperature reach the setting temperature 220°C. Then 1 metric ton pressure was applied for a few minutes. The hot press was shut off to cool down while keeping the pressure till it reached 80°C, then the pressure was released completely from the mold except the weight of the upper aluminum plate and let it cooled down to the room temperature. The CNT yarns were embedded into the PMMA matrix to form PMMA/CNT-yarn composite through the hot-press process. The thickness of the composite sheets was controlled between 20-30 μm. The CNT content in the composite was controlled by the number of the CNT yarns.

Property characterization: Optical microscope was used to observe the CNT yarn arrays in PMMA matrix. The detail structures of the CNT yarns and PMMA/CNT-yarn composite were also observed using scanning electron microscope. For evaluating the content of the CNTs in composite and the thermal stability of the composite, Thermogravimetric Analysis was conducted on PMMA and PMMA/CNT composites in nitrogen from 25 °C to 450 °C and then in air from 450 °C to 800 °C at a heating rate of 10 °C/min. The tensile samples were cut from the hot-pressed PMMA/CNT-yarn composite. The engineering tensile stress-strain behaviors of PMMA sheet, CNT yarns, and the PMMA/CNT-yarn composites were tested by using Dynamic Mechanical Analysis (DMA). The gauge length, film width, and force rate for the tensile test was ~10 mm, 3~4 mm, and 0.5 N/min or 1.0 N/min, respectively. The loading direction is parallel to the direction of CNT yarn alignment.

3. Results and Discussion

The CNT yarns were twist spun from the CNT forest. The CNTs are multi-walled carbon nanotubes with ~10 nm diameter and 400-500 μm length. The CNTs in a yarn are well aligned with an orientation angle to its axis. The diameter of the yarns is between 10-20 μm and the twist angle is around 30 degree.

---

9 . M. Zhang, K. R. Atkinson, and R. H. Baughman, "Multifunctional Carbon Nanotube Yarns by Downsizing an Ancient Technology," *Science* **306**, 1358 (2004).

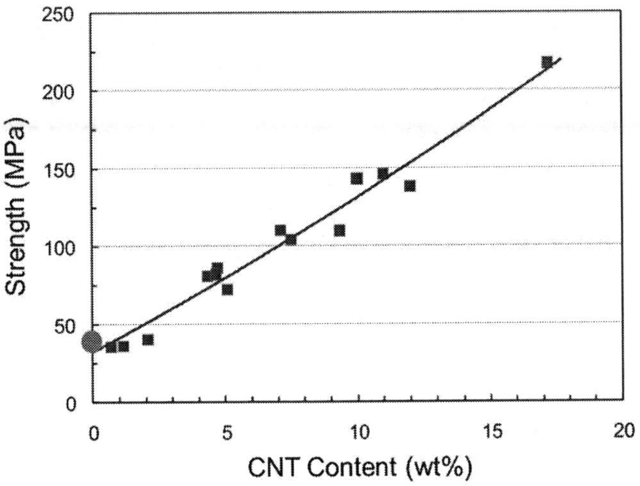

Figure 1. Effect of CNT content on strength of the PMMA/CNT-yarn composites. Red dot in the graph shows the data of the neat PMMA sheet.

The PMMA/CNT-yarn composites were made by applying CNT yarns on top of a PMMA sheet to form a CNT yarn array, and then the array was embedded into the PMMA matrix through the hot-press process. The CNT yarns are straight and uniformly distributed in the composite. This approach allows us to form well designed CNT networks inside polymer matrices and it meets the fundamental requirements for filler: high aspect ratio, good distribution in polymer matrix, good alignment, and better interfacial stress transfer.

The engineering stress-strain behaviors of neat PMMA sheets, CNT yarns, and PMMA/CNT-yarn composites were tested by DMA under constant loading rate. The tensile strength and strain of the neat PMMA sheet are ~40 MPa and 2.7% respectively. The pristine CNT yarn has about 10 times higher strength and 3.5 times longer strain than that of the neat PMMA sheet. The PMMA/CNT-yarn composites show dramatic increase in tensile strength and modulus with the increase of CNT content. The change of the strength of the composite with the CNT content is shown in Fig. 1.

Compared to the neat PMMA sheets, the increase of CNT content in the composite results in the increase of the strength. In addition to the increase in strength, the increase of the CNT content also leads to a significant increase of the modulus of the composite. The PMMA/CNT-yarn composite with 10 wt% CNT yarns shows an increase of 250% and 500% in tensile strength and modulus respectively with a little deduction in elongation. When the CNT content increases to 17 wt%, the composite has 5 fold increase in strength and 10 fold increases in modulus. Increases in elastic modulus and ultimate strength indicate that nanotubes are acting as reinforcement in the polymer matrix by directly transferring load from nanotube to nanotube and from the polymer to the nanotubes. The effective enhancement is because that the CNTs in the composite are in the highly organized structure and the CNTs are oriented close to the loading direction.

4. Conclusions

CNTs were assembled into highly ordered structure, CNT yarns and CNT-yarn array. By using CNT-yarn array as filler, the effect of CNT reinforcement to the polymeric matrix is demonstrated. We find that CNT-yarn array effectively enhance the mechanical properties of the PMMA matrix. The modulus and strength of the PMMA/CNT-yarn composite increase by a factor of 10 and 5, respectively at 17 wt% CNT content. We believe that the effective reinforcement is due to the highly ordered CNT structure and the CNT content in the polymer matrix. To further improve the properties of the CNT yarn and/or to increase the content of CNTs could lead to even better reinforcement effects. Except for the enhancement in mechanical properties, CNT yarns in composite can also provide high electrical and thermal conductivities for various applications. This technology can be applied to other transparent and non-transparent polymeric, metal, and ceramic materials.

Acknowledgments

This work is supported by AFRL.

(333b) HRTEM & XPS Applied to Particulate Emissions

Randy L. Vander Wal, Chung-Hsuan Hunag and Jane Hitomi Fujiyama Novak, Energy and Mineral Engineering & The EMS Energy Institute, Pennsylvania State University, University Park, PA

Characterization of source (particulate) emission profiles is integral to formulation of source profiles towards development of receptor-based models and extraction of source contributions to monitored sites. Exposure and health effects can then be accurately evaluated, atmospheric impacts accurately assessed and regulatory statutes specifically targeted.

Electron microscopic and spectroscopic methods have great potential for carbonaceous aerosol characterization. High-resolution transmission electron microscopy (HRTEM) provides information of the carbon nanostructure (or relative lack thereof). Image quantification allows extraction of various statistical parameters including lamella length, separation distance and tortuosity. These may be summarized in the form of histograms. X-ray photoelectron spectroscopy (XPS) provides a means for probing the surface chemistry of materials. It provides not only information about the atomic composition of a sample but also information about the structure and oxidation state of the constituent elements.

Combustion produced soot is a highly variable material. Physically the nanostructure can range from amorphous to graphitic to fullerenic. Chemically nearly any element could be included, though the surface functional groups are predominantly oxygen-based. Results will be presented for analysis of HRTEM images of the physical nanostructure and XPS analysis of the chemical composition of soots collected from plant, institutional and residential scale oil-fired boilers, diesel engine, gas-turbine combustor (jet engines) and a wildfire. Physically soots from these emission source classes may be differentiated based on carbon lamella length, mean separation and tortuosity. Chemically these soots may also be distinguished by elemental composition, surface (oxygen) functional groups and carbon bonding.

Characterization of source (particulate) emission profiles is integral to formulation of source profiles towards development of receptor-based models and extraction of source contributions to monitored sites. Exposure and health effects can then be accurately evaluated, atmospheric impacts accurately assessed and regulatory statutes specifically targeted.

See more of this Session: Green and Renewable Energy Aspects of Nanotechnology
See more of this Group/Topical: Environmental Division

(333c) Nanostructure Engineering of Polymer Solar Cells

Thomas Mensah, Georgia Aerospace Corporation, Atlanta, GA

Energy and environment have become two of the top concerns and challenges of the human beings nowadays, and the two issues are closely related to each other. An estimated over 80% of the world energy supplies are from the burning of fossil fuels such as coal, oil, and natural gas [1]. However, carbon dioxide and other toxic gases released from fossil fuel burning are causing many actual and potential environmental problems such as air pollution and globe warming. In addition, fossil fuel deposits on the earth are not unlimited. Sunlight is perhaps the most abundant, free, renewable, and clean energy source on the earth. While the worldwide energy power consumption in 2006 is less than 20 Terawatts, sunlight provides about 165000 Terawatts power to the earth. Though the commercially available inorganic materials based solar cells have typical photoelectric power conversion efficiencies of over 10%, and a five-year compound annual growth rate (CAGR) of 56% through 2008 of solar cell industry is impressive, the technology is still too expensive for popular or general power supplies. Future generation organic or 'plastic' materials based solar cells appear very attractive for popular solar energy conversion applications where low cost (such as large area), lightweight, and flexible shape are desired. The best photoelectric power conversion efficiencies of currently reported organic/polymer photovoltaic materials is still relatively low (about 5-7%), but there is a large room for improvements. This presentation will first provide an overview on the history, fundamental mechanisms, and current status of organic/polymeric photovoltaic materials and devices [2]. A number of approaches including approaches pursued at the speaker's lab will be presented. Prospective and ways of further improving organic/polymeric solar cells will also be briefly addressed. [1] Sun, S and Sariciftci, N. eds., Organic Photovoltaics: Mechanisms, Materials and Devices (ISBN 0-8247-5963-X). CRC Press/Taylor Francis, Boca Raton, Florida, USA, 2005. [2] Sun, S. "Organic and Polymeric Solar Cells", in Handbook of Organic Electronics and Photonics, edited by S. H. Nalwa, American Scientific Publishers, Los Angeles, California, 2006, vol. 3, chapter 7, pp 314-350.

See more of this Session: Green and Renewable Energy Aspects of Nanotechnology
See more of this Group/Topical: Environmental Division

SILICON-COATED CARBON NANOTUBE ANODES FOR LITHIUM-ION BATTERIES

Dr. Michelle K. Gaines, Georgia Tech Research Institute – Electro-Optical Systems Laboratory, Atlanta, GA 30332

Dr. Edward D. Sosa, NASA – Lyndon B. Johnson Space Center, Houston, TX 77058

Alexandre L. Magasinski, Georgia Institute of Technology – School of Materials Science & Engineering, Atlanta, GA 30332

Samuel K. Karpowicz, Georgia Institute of Technology – School of Mechanical Engineering, Atlanta, GA 30332

Deborah S. Williams, School for Technology, Engineering, Math, and Science at Therrell, Atlanta, GA 30311

Introduction

As global industries refocus their efforts to reduce harmful emissions and greenhouse gases by moving away from petroleum based fuel, electric vehicles and hybrid-electric vehicles are being closely considered as an attractive replacement. Before such technologies move to market, the energy storage industry must further mature in order to reliably power these large devices [1]. Lithium-ion batteries have moved to the forefront of rechargeable batteries because they have a high energy storage capacity, relative to its weight, which provide an economical solution for powering portable electronic devices [1-4]. NASA is particularly interested in lightweight power sources, which are high power and have higher energy densities. Powerful lightweight rechargeable batteries will be essential in meeting the energy requirements for Lunar and Mars missions, as well as outer planet exploration. Since Sony brought lithium-ion batteries to the market in the early 90s, many materials have been introduced to assist the electrochemical activity for lithium-ion exchange. The cathode has been primarily composed of spinel oxides (such as $LiCoO_2$ or $MnCo_2O_4$) for their ability to catalyze oxygen reduction reactions [5]. Graphite has most often been used for the anode because of its unique ability to interact with ions through intercalation of its graphene sheets [2-5]. This allows for intimate contact with the lithium ions to engage in electrochemical reactions.

In order to reliably power large electronic devices and vehicles built for spacecraft within the next 10-15 years, other materials for lithium-ion batteries must be considered [1, 6-8]. The energy storage capacity must greatly increase. The spinel oxides have still been identified as sufficient cathode materials; however there is much room for development of the anode materials [7-10]. The theoretical storage capacity of graphite is 372 mAh/g. Silicon has a theoretical storage capacity of 4200 mAh/g, which is an order of magnitude higher. The problem with silicon is its breakdown in microstructure after Li^+ cycling. Although silicon's most lithiated alloy is $Li_{22}Si_5$, proving that silicon is able to store far more Li^+ than graphite (LiC_6), its structure doesn't accommodate the Li^+s the same way graphite does [8, 11]. The extreme volume changes due to Li^+ alloying and dealloying (280% – 420%) causes its structure to disintegrate [8-9, 11-17].

Many solutions have been investigated for preventing silicon anode capacity fade due to extreme volume expansion. Some of these include creating silicon-carbide alloys out of carbide-derived carbon (CDC) and silicon-graphite blends [12, 15]. Others include: 1) coating carbon nanoparticles with thin silicon films [8, 13], 2) growing pure silicon nanowires onto metal substrates, leaving the nanowires as the active material [9, 16], and 3) coating fullerene thin films onto silicon [17].

This project focused on creating Li^+ anodes of silicon-decorated carbon nanotube (CNT) forests on copper current collectors (Figure 1). The anodes were assembled into half 2032-coin cells, for identifying their electrical performance, in an effort to retain the silicon active material for at least 100 galvanic cycles.

Figure 1. Illustration of Li^+ charge/discharge process of Si-coated CNTs on copper substrates.

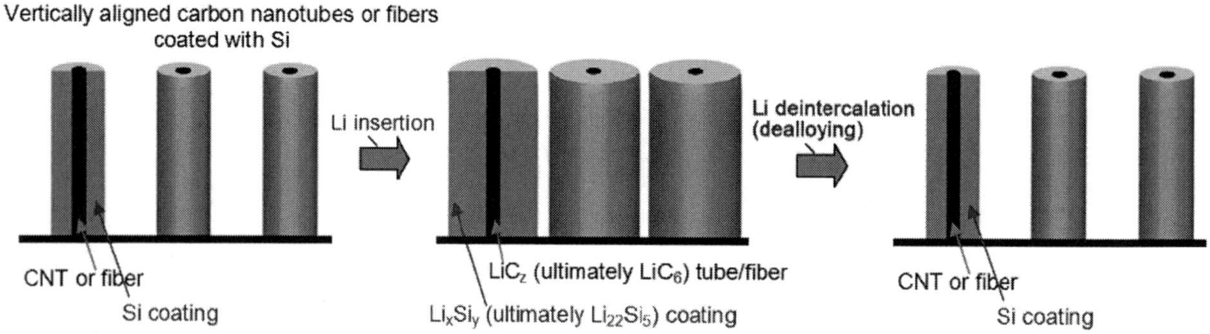

The CNTs are designed to serve as mechanical reinforcements for the silicon, to maintain the active material's microstructure after undergoing large volume changes from Li^+ charging and discharging.

Experimental

CNT Growth on Copper

The project began with the non-trivial task of growing CNTs directly onto copper substrates. Two methods were employed: 1) Annealing a 10wt% solution of iron(III) *p*-toluenesulfonate salt (Alfa Aesar) and ethanol, causing iron catalyst nanoparticles to form on the surface to seed thermally-enhanced chemical vapor deposition (TECVD) for CNT growth, and 2) Depositing a series of metal diffusion layers (Ti/ Ni/ Ti/ Al) between the copper surface and a 10 nm layer of iron to seed plasma-enhanced chemical vapor deposition (PECVD) for CNT growth. The iron salt substrates were placed into a chemical vapor deposition (CVD) furnace (FirstNano EasyTube) for an anneal step (slow ramp to 600°C under Ar atmosphere), an Fe(III) reduction to Fe nanoparticles step (600°C for 30 min under H_2 and Ar), and a CNT deposition step (800°C for 10 min or 10 min C_2H_2 gas, H_2, and Ar) [18]. The PECVD substrates were placed into a Black Magic (Aixtron) furnace for a 10 minute anneal step under NH_3 gas and 80W, 15kHz, and 725V plasma, and a 10-minute CNT deposition step under C_2H_2 gas at 725°C.

Silicon Deposition

The next step in the project focused on depositing silicon thin films onto the surface of the CNTs. The Cu-CNT samples were placed in a RF-PECVD reactor (Unaxis 790 Series) under the following conditions: 13.56 MHz, radio-frequency power 20 mW/cm^2, 500 sccm SiH_4 gas (5% He), under 400 mTorr and 200°C. The deposition rate of the tool was 80 Å/min, and the samples were subjected to either a 2-minute or 8-minute deposition time interval (Si wall thickness = 16nm and 64 nm, respectively).

Electrical Measurements

Electrodes were fabricated and assembled into 2032- (stainless steel) coin cells. The Cu-CNT-Si samples (~0.002 inch thick) were punched out into ½ inch circles. For improved electrical contact, the electrodes were spot-welded to a spacer and bottom of the coin cell [14]. Two-electrode half-cells were fabricated with Li foil (0.03 inch thick) (Alfa Aesar) as the counter and reference electrode, and with 25 μm of polypropylene/polyethylene/polypropylene trilayer separator (Celgard 2325) soaked in 1M LiPF6 salt solution with ethylene carbonate:diethyl carbonate:dimethyl carbonate (Alfa Aesar) (1:1:1 by volume) electrolyte [16]. The cell assembly was carried out in an argon filled glove box. Galvanic cycling and electronic impedance spectroscopy measurements were performed using a SI1286 galvanostat/ potentiostat (Solartron) and analyzed on CorrWare.

Results and Discussion

CNT Morphology

Figure 2a shows scanning electron microscope (SEM) images of multi-walled carbon nanotubes (MWCNT) grown on copper via TECVD. Figure 2b illustrates MWCNTs grown via PECVD.

Figure 2. CNTs grown via (a) TECVD for 30 minutes and (b) PECVD for 10 minutes.

(a) (b)

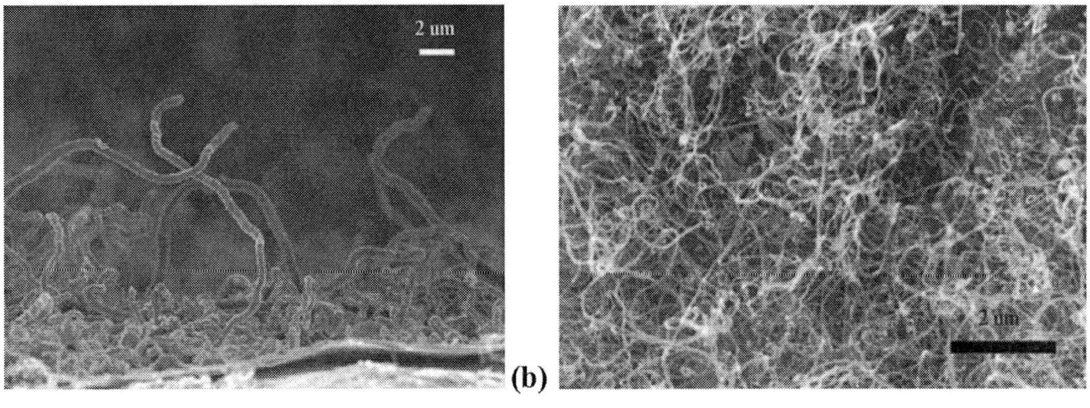

Fig. 2a shows a carpet of MWCNTs grown onto the substrate via TECVD for 30 minutes at 800°C. Fig. 2b shows the result of PECVD for a 10 minute deposition time interval at 725°C. Some of the advantages of plasma-enhancement over thermal-enhancement is the ability to grow at lower temperatures and with better control over CNT morphology [19]. Although both images show the MWCNT looking like spaghetti on the surface of the copper, they depict two different nanotube diameters. Fig. 2a has much wider tubes of about 500 nm, and Fig. 2b has much narrower tubes at about 100 nm.

In Fig. 2a, the variation in CNT length, diameter, and lack of vertical-alignment is the result of several factors: 1) variation of catalyst topography and density due to catalyst spin casting, 2) poor catalyst adhesion, 3) TECVD, and 4) less than ideal parameters for deposition gas flowrate. The catalyst salt has a much higher surface roughness than a pure metal film, electron-beam evaporated onto a surface (Fig. 2b). This increased surface roughness will cause more hindrance in the reduction reaction kinetics from Fe(III) to Fe nanoparticles and will not produce a uniform layer of Fe nanoparticles. Catalyst delamitation on the copper substrate is evidenced at the bottom of 2a, a common hurtle when depositing diffusion layers on to a metal substrate. Delamination was avoided (as shown in Fig. 2b) by electron-beam evaporating Ti, Ni, Ti, and Al layers at various thicknesses and by depositing the CNTs under plasma-enhancement under lower temperatures. Although TECVD provides enough energy to carry out the deposition reactions, PECVD gives a better result [20]. Other labs have highlighted the importance of uniform particle size and narrow nanotubes for increased electrochemical performance [7, 16, 21-22]. Developing narrower, shorter, uniform CNTs is a goal for this research, which will deliver improved electrode performance.

Silicon Deposition & Characterization

Figure 3 illustrates a map of the silicon coating on the surface of the CNTs via electron-backscattering detection on an S-4700 SEM.

Figure 3. SEM electron-backscattering show more contrast between deposited silicon on top of the surface of the CNTs. (a) Fe(III) salt anodes and (b) Ti/ Ni/ Ti/ Al diffusion layered anodes. The bright spots represent the presence of silicon.

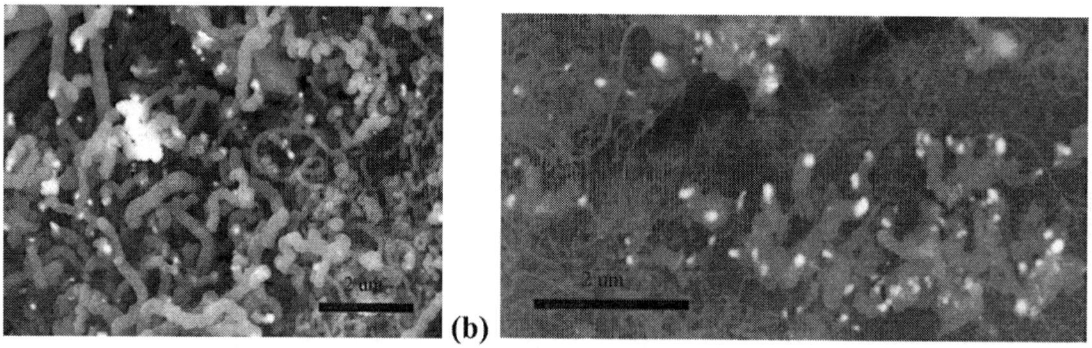

(a) (b)

The electron-backscattering detector reveals that only the tips of the CNTs were coated with silicon and the bases, near the Cu substrate, were not. These images are similar to what's been reported for silicon deposition on the surface MWCNT [23].

Figure 4 demonstrates via Raman spectroscopy that both amorphous and nanocrystalline silicon are present. Measurements were collected at three different locations on an Fe(III) salt sample (as imaged in Fig. 2a) at 785 nm, 633 nm, and 514 nm laser excitation. All three locations depict a large, broad amorphous silicon peak at about 480 cm^{-1}. One of the measurements reveals a second short, sharp peak at about 525 cm^{-1}, which is indicative of nanocrystalline silicon.

Figure 4. Raman spectroscopy performed at 785 nm laser excitation on a Fe(III) salt sample.

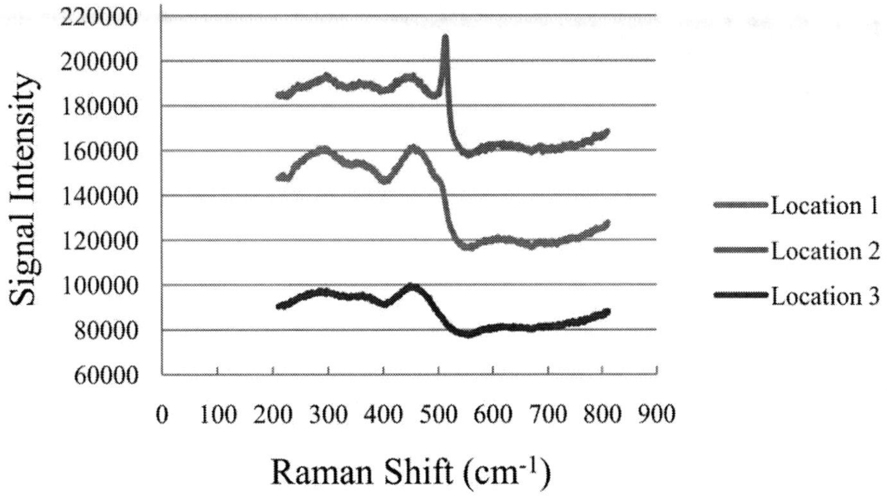

The results highlighted in Fig. 4 agree well with result reported upon by Nguyen and coworkers. They explain that the chamber pressure (400 mTorr), silane flowrate (500 sccm), and temperature (200°C), cause the deposited silicon to fall in the α-regime, producing a less dense plasma from high RF power or pressure [23]. Silicon in the α–regime has a lower particle density and behaves more capacitive than resistive, which would serve as a benefit for producing a better performing battery. If the pressure were at least 1000 mTorr, the silicon would be in the γ-regime and have higher particle densities and behave more resistive [23]. A challenge is to deposit silicon along the entire surface of the MWCNT, all the way to the Cu substrate. Previous work has highlighted the benefit of direct contact between the active material and the current collector [8-9, 13, 15, 24-25]. Nguyen and coworkers report that less dense CNT carpets will allow for more molecular silane gas transport to the surface of the substrates, allowing for adequate silicon coating of CNTs. X-ray photoelectron spectroscopy (XPS) measurements were also performed on the surface of the Fe(III) anodes. Although not pictured, they showed that silicon was present on the CNT surface.

Electrical Characterization

Several electrical measurements were performed on the Fe(III) salt 2032- half cell samples. The experiments performed were electrical impedance spectroscopy (EIS), galvanic cycling, and cyclic voltammetry (CV). EIS and CV measurements are not shown here, but the EIS measurements illustrated both a surface resistance kinetic contribution as well as a diffusion contribution, due to the Li^+s moving into the bulk Si-CNT active material, creating surface resistance from ionic conduction through the electrolyte the and electron transfer resistance on the electrode interface and the solid electrolyte interface (SEI). The CV curves showed a typical profile for Li^+ extraction and insertion.

Galvanic cycling experiments were performed under a potential range of $0.01 \text{ V} < E < 1.0 \text{ V}$ and 0.1 mA of current. First, Li^+s were discharged and charged twice under a slow charge/discharge rate of *C/20*, which establishes the SEI. Secondly, the cells were charged and discharged at a fast rate of *C/2* for 100 cycles and then cycled twice at a slow rate of *C/20*, respectively. The specific storage capacity as a function of cycle number is illustrated for one of the Fe(III) salt samples in Figure 5.

Figure 5. Specific storage capacity (mAh/g) as a function of cycle number, reported from galvanic cycling experiments performed on an Fe(III) salt anode.

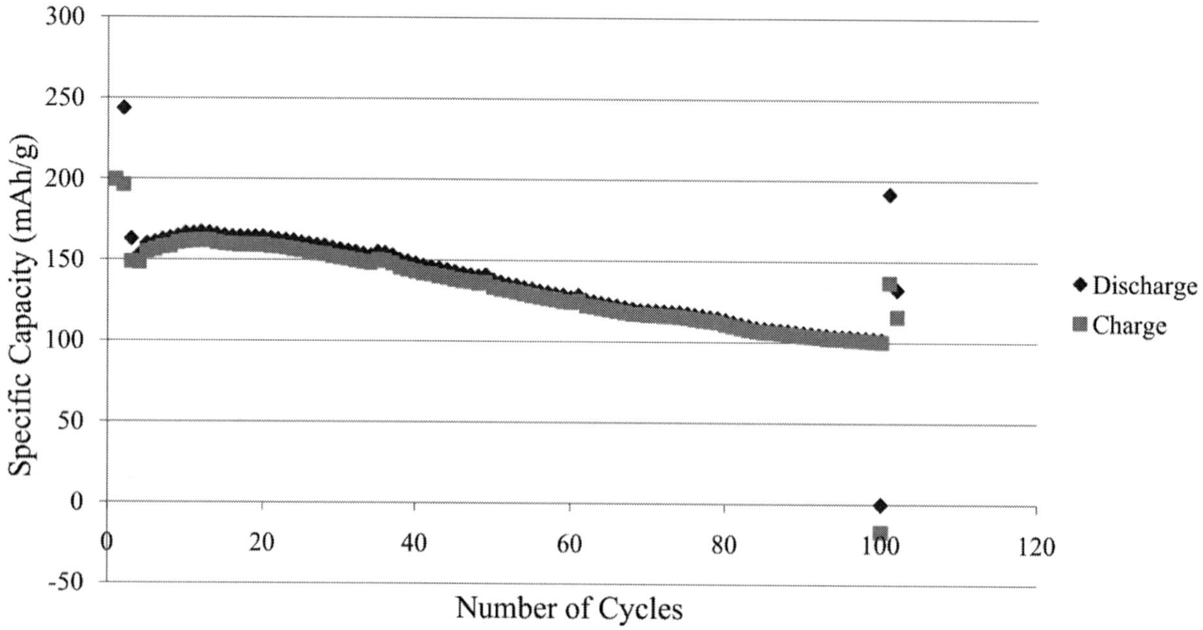

Fig. 5 illustrates the charge and discharge specific storage capacities measured on one of the Fe(III) salt samples. The columbic efficiency between the measured specific capacity during charging and discharging was about 98%, evidencing strong consistency between the charging and discharging measurements. The Li^+ storage capacity ranged from a maximum of 160 mAh/g to a minimum of 100 mAh/g. Although these are very low storage capacities for silicon, these results are very accurate to what has been reported upon for Li^+ storage capacity of plain CNT anodes, reflecting that most of the active material present are uncoated CNTs. The results displayed in Fig. 5 also agree well with the backscattered SEM images, which mapped a very small amount of silicon with respect to the CNTs. The results illustrated in Fig. 5 also reveal that the anode structure is maintained over the course of the 100 cycles, without a large drop in specific capacity.

Conclusions

Lithium-ion batteries composed of silicon-decorated CNT anodes have been developed and fabricated. There have been some demonstrated successes, including growing the CNTs directly onto the copper current collector, depositing amorphous and nanocrystalline silicon thin films on to the CNTs' surface, and demonstrating consistent Li^+ storage capacities for at least 100 cycles from galvanic cycling experiments.

Acknowledgements

This work was funded by the NASA Johnson Space Center, contract # NNX10AG09G. We would like to acknowledge Dr. Samuel Graham and Olalekan Adewuyi from Georgia Tech's School of Mechanical Engineering for their contribution towards suggesting the four-metal diffusion layer method for achieving CNT growth on copper. We would also like to acknowledge Dr. Gleb Yushin in Georgia Tech's School of Materials Science and Engineering for their provision of coin cell fabrication and the electrical measurements. S.K.K. and D.S.W. are grateful for their funding from the Summer Undergraduate Research in Engineering & Science (SURE) and the Georgia Internship Fellowship for Teachers (GIFT) programs.

References:

1. Burke, A.F., *Batteries and Ultracapacitors for Electric, Hybrid, and Fuel Cell Vehicles.* Proceedings of the IEEE, 2007. **95**(4): p. 806-820.
2. Noel, M. and V. Suryanarayanan, *Role of carbon host lattices in Li-ion intercalation/de-intercalation processes.* Journal of Power Sources, 2002. **111**(2): p. 193-209.
3. Sethuraman, V.A., et al., *Surface structural disordering in graphite upon lithium intercalation/deintercalation.* Journal of Power Sources, 2010. **195**(11): p. 3655-3660.
4. Hu, Y.-S., et al., *Synthesis of Hierarchically Porous Carbon Monoliths with Highly Ordered Microstructure and Their Application in Rechargeable Lithium Batteries with High-Rate Capability.* Advanced Functional Materials, 2007. **17**(12): p. 1873-1878.
5. Sugawara, M., M. Ohno, and K. Matsuki, *Oxygen reduction catalysis of Mn-Co spinel oxides on a graphite electrode in alkaline solution.* Journal of Materials Chemistry, 1997. **7**(5): p. 833-836.
6. Yan, J., et al., *Preparation and electrochemical properties of composites of carbon nanotubes loaded with Ag and TiO2 nanoparticle for use as anode material in lithium-ion batteries.* Electrochimica Acta, 2008. **53**(22): p. 6351-6355.
7. Landi, B.J., et al., *Multi-Walled Carbon Nanotube Paper Anodes for Lithium Ion Batteries.* Journal of Nanoscience and Nanotechnology, 2009. **9**(6): p. 3406-3410.
8. Wang, W. and P.N. Kumta, *Nanostructured Hybrid Silicon/Carbon Nanotube Heterostructures: Reversible High-Capacity Lithium-Ion Anodes.* Acs Nano, 2010. **4**(4): p. 2233-2241.
9. Kuksenko, S.P., *Cycling Parameters of Silicon Anode Materials for Lithium-Ion Batteries.* Russian Journal of Applied Chemistry, 2010. **83**(4): p. 641-647.
10. Reddy, A.L.M., et al., *Coaxial MnO2/carbon nanotube array electrodes for high-performance lithium batteries.* Nano Letters, 2009. **9**(Compendex): p. 1002-1006.
11. Shenoy, V.B., P. Johari, and Y. Qi, *Elastic softening of amorphous and crystalline Li-Si Phases with increasing Li concentration: A first-principles study.* Journal of Power Sources, 2010. **195**(19): p. 6825-6830.
12. Jo, Y.N., et al., *Si-graphite composites as anode materials for lithium secondary batteries.* Journal of Power Sources, 2010. **195**(18): p. 6031-6036.
13. Xie, J., G.S. Cao, and X.B. Zhao, *Electrochemical performances of Si-coated MCMB as anode material in lithium-ion cells.* Materials Chemistry and Physics, 2004. **88**(2-3): p. 295-299.

14. Magasinski, A., et al., *High-performance lithium-ion anodes using a hierarchical bottom-up approach.* Nat Mater, 2010. **9**(4): p. 353-358.
15. Portet, C., G. Yushin, and Y. Gogotsi, *Effect of Carbon Particle Size on Electrochemical Performance of EDLC.* Journal of The Electrochemical Society, 2008. **155**(7): p. A531-A536.
16. Ruffo, R., et al., *Impedance Analysis of Silicon Nanowire Lithium Ion Battery Anodes.* The Journal of Physical Chemistry C, 2009. **113**(26): p. 11390-11398.
17. Arie, A.A. and J.K. Lee, *A study of Li-ion diffusion kinetics in the fullerene-coated Si anodes of lithium ion batteries.* Physica Scripta, 2010. **T139**.
18. Chen, J., et al., *Direct Growth of Flexible Carbon Nanotube Electrodes.* Advanced Materials, 2008. **20**(3): p. 566-570.
19. Bell, M.S., K.B.K. Teo, and W.I. Milne, *Factors determining properties of multi-walled carbon nanotubes/fibres deposited by PECVD.* Journal of Physics D (Applied Physics), 2007. **40**(Copyright 2007, The Institution of Engineering and Technology): p. 2285-92.
20. Bell, M.S., et al., *Characterisation of the growth mechanism during PECVD of multiwalled carbon nanotubes*, in *Carbon. The Future Material for Advanced Technology Applications*. 2006, Springer-Verlag: Berlin, Germany. p. 77-93.
21. Chmiola, J., et al. *Electric double-layer capacitance of multi-wall carbon nanotubes and carbide derived carbon.* in *207th Meeting of the Electrochemical Society, May 15, 2005 - May 20, 2005*. 2005. Quebec, Canada: Electrochemical Society Inc.
22. Chmiola, J., et al., *Effect of pore size and surface area of carbide derived carbons on specific capacitance.* Journal of Power Sources, 2006. **158**(Compendex): p. 765-772.
23. Nguyen, J.J., K. Evanoff, and J.W. Ready, *Amorphous and Nanocrystalline Silicon Growth on Carbon Nanotube Substrates.* manuscript in preperation for submission to Advanced Materials, 2010.
24. Portet, C., G. Yushin, and Y. Gogotsi, *Electrochemical performance of carbon onions, nanodiamonds, carbon black and multiwalled nanotubes in electrical double layer capacitors.* Carbon, 2007. **45**(13): p. 2511-2518.
25. Landi, B.J., et al., *Carbon nanotubes for lithium ion batteries.* Energy & Environmental Science, 2009. **2**(6): p. 638-654.

(333e) Flow Behavior, Morphology and Properties of Multiscale Polymer Nanocomposites

Thomas Mensah, Georgia Aerospace Corporation, Atlanta, GA

Nanoengineered composite materials, in which one or more of the composite constituents (eg., matrix, fibers, and laminates) integrates carbon nanotubes and possesses enhanced and functional properties, represent a potentially enabling technology for multifunctional composite materials. Optimization of the processing and resulting properties of these advanced materials is a significant challenge. While a number of studies have reported on polymer/CNT nanocomposites in which the tubes are randomly dispersed in the matrix, controlled spatial distribution of the CNTs provides more control over properties and opens the door to an even wider range of applications. However, control of the dispersion is very challenging. While many studies have investigated relationships between various processing methods, the resulting dispersion and the properties, we have found that state of nanoparticle dispersion is intimately linked to the viscoelastic properties of the nanodispersed resin and its interaction with the process flow field. In the present study, we have investigated these relationships using rheology of the nanofluids and morphological analysis of the composites. We have used our findings to prepare fiber-reinforced composite laminates. The dispersion state and properties will be presented.

See more of this Session: Green and Renewable Energy Aspects of Nanotechnology
See more of this Group/Topical: Environmental Division

(333f) Three Dimensional Carbon Nanotube Photovoltaics

Jack D. Flicker, Electro-optical System Lab, Georgia Tech Research Institute, Atlanta, GA and W. Jud Ready, Electro-Optical System Lab, Georgia Tech Research Institute, Atlanta, GA

Novel solar cell architectures, which increase power output compared to traditional two dimensional designs, are desirable for the coming years. We introduce a three dimensional photovoltaic device with vertically aligned carbon nanotube pillars coated with photoactive materials to create a solar cell. The extra dimensionality of this cell, as added by the nanotubes, has been theorized to increase the energy generated over planar cells by up to four times. This energy increase is due to an increase in the interactions between photons and the photoactive material as the sun is at an off-normal angle to the cell substrate. This allows photons to "pinball" between vertical faces of the carbon nanotube towers, greatly increasing the chances that a photon is absorbed rather than reflected away from the cell face. Prototypes of these cells have been made and, when the solar flux is at an off normal angle to the substrate, do show an increased power production much as theory predicts. Current work focuses on further characterizing this increase in power output over two dimensional cells and increasing the overall efficiency of these cells through processing parameters.

See more of this Session: Green and Renewable Energy Aspects of Nanotechnology
See more of this Group/Topical: Environmental Division

(333g) Synthesis of Fe-Based Catalysts Coupled with Carbon Mineral Sequestration

Xiaozhou Helios Zhou and Ah-Hyung Alissa Park, Earth and Environmental Engineering & Chemical Engineering, Columbia University, New York, NY

Exploring clean and inexpensive energy and tackling climate change are two of the most pressing issues being faced in the 21st century. Conventional fossil fuel energy especially oil features unstable and periodically soaring price along with significant CO2 emission. Therefore, the missions to deal with global energy and climate constraints are coupled more than ever. One of the methods to store CO2 permanently is carbon mineralization. This method is achieved by chemically fixing gaseous CO2 into solid mineral matrix as carbonates, which is argued to be the safest and most permanent method of carbon storage to date. Unfortunately, most of the carbon mineralization processes are energy intensive and expensive. Thus, there have been significant efforts to reduce the cost of carbon mineral sequestration technology. One of the useful byproducts of carbon mineralization is iron oxide. Silicate minerals such as serpentine often contain 5-10 wt% of iron. Since iron is extracted into aqueous phase during the carbon mineralization process, a variety of Fe-based materials including iron oxide nanoparticles can be synthesized. In this study, iron-based catalysts are synthesized from serpentine and evaluated for two applications: water-gas-shift and biomass-to-hydrogen reactions. Since solutions containing Fe are generated in the presence of chelating agents targeting Mg in serpentine, the effects of chelating agents, alkalizing agents, ionic strength, pH, temperature and post treatment on the synthesis of iron-based catalysts are investigated. Eventually, the utilization of these iron oxide catalysts would further improve the carbon sequestration capacity of the mineral carbonation technology.

See more of this Session: Green and Renewable Energy Aspects of Nanotechnology
See more of this Group/Topical: Environmental Division

(413a) Oriented Single-Crystalline Rutile TiO2 Nanorods On Transparent Conducting Substrates for Dye-Sensitized Solar Cells

Bin Liu, Emil Enache-Pommer and Eray S. Aydil, Department of Chemical Engineering and Materials Science, University of Minnesota, Minneapolis, MN

Dye-sensitized solar cells (DSSCs) made from oriented, one-dimensional (1D) semiconductor nanostructures such as nanorods, nanowires, and nanotubes are receiving wide attention because direct connection of the point of photogeneration with the collection electrode using such 1D structures may improve the cell performance. Specifically, oriented single-crystalline TiO2 nanorods or nanowires on a transparent conductive substrate would be most desirable, but achieving these structures has been limited by the availability of synthetic techniques. In this presentation, we describe a facile hydrothermal method to grow oriented, single-crystalline rutile TiO2 nanorod films on transparent conductive fluorine-doped tin dioxide (FTO) substrates. The diameter, length, and density of the nanorods could be varied by changing the growth parameters such as growth time, growth temperature, initial reactant concentration, acidity and additives. The epitaxial relation between the FTO substrate and rutile TiO2 with a small lattice mismatch plays a key role in driving the nucleation and the subsequent growth of the rutile TiO2 nanorods. DSSCs were assembled using the TiO2 nanorods grown on FTO as the photoanode. A light-to-electricity conversion efficiency of 3% could be achieved by using 4 μm-long TiCl4-treated TiO2 nanorod films. We further compared the electron transport and recombination rates in these single-crystal rutile TiO2 nanorods with TiO2 nanoparticle films of similar thickness. Surprisingly, we find that the electron transport time constant is approximately the same and even slightly slower in single-crystal rutile TiO2 nanorods than in TiO2 nanoparticle films, suggesting that electron diffusion rate is still determined by the residence time in surface traps even in single-crystal TiO2 nanorods.

See more of this Session: Nanomaterials for Photovoltaics I
See more of this Group/Topical: Topical 5: Nanomaterials for Energy Applications

(413b) Sub-Picosecond Dynamics of Free Electrons in TiO_2 Nanotubes

Christiaan Richter, Diyar Talbayev and Charles A. Schmuttenmaer, Department of Chemistry, Yale University, New Haven, CT

The sub-picosecond mobility and lifetime of free electrons injected into TiO_2 nanotubes were probed with sub-picosecond temporal resolution using time resolved terahertz spectroscopy (TRTS). The photoconductivity spectra of the nanotubes revealed the free electrons to be very mobile, however most electrons appear to be trapped by a single kind of defect present at high concentration (> 10^{14} cm^3) having a distinct resonance at 1.8 THz (7.5 meV). Synthesizing the nanotubes on the walls of a waveguide allows for probing the shallow defect structure of the nanotubes with even greater sensitivity than is possible in transmission though free space. Insights obtained by TRTS indicate the potential, and also reveal the challenges, of using TiO_2 nanotubes and other semiconductor nanostructured materials as charge transport conduits in photovoltaic cells.

See more of this Session: Nanomaterials for Photovoltaics I
See more of this Group/Topical: Topical 5: Nanomaterials for Energy Applications

(413c) Long TiO2 Nanotube Arrays Synthesized Directly On Transparent Conducting Oxide for Efficient Dye-Sensitized Solar Cells

Chengkun Xu, Chemical Engineering, University of Pittsburgh, Pittsburgh, PA and Di Gao, Chemical and Petroleum Engineering, University of Pittsburgh, Pittsburgh, PA

Vertically aligned TiO2 1-D nanostructures synthesized directly on transparent conducting oxide (TCO) have shown great promise to overcome the limitation of the random network of TiO2 nanoparticles currently used in dye-sensitized solar cells (DSCs). Synthesis of such structures with a sufficient length for fabrication of high-efficiency DSCs, however, remains challenging. Here we show that by using a wet chemical process, arrays of TiO2 nanotubes up to 20 μm long can be directly synthesized on TCO through conversion of ZnO nanowire arrays. DSCs fabricated using the synthesized TiO2 nanotube arrays yield a power conversion efficiency of 6.1%, which is comparable to TiO2 nanoparticle-based DSCs fabricated using the same dye and electrolyte. In addition, the anode film as thick as 21 μm based on such TiO2 nanotube arrays possess a high electron collecting efficiency and an appreciable enhancement in the red light quantum yield is observed on the DSCs with such anodes. This result suggests numerous opportunities for further improvement of DSCs, for example, by employing solid-state electrolytes and redox mediators with faster kinetics.

See more of this Session: Nanomaterials for Photovoltaics I
See more of this Group/Topical: Topical 5: Nanomaterials for Energy Applications

(413d) Modeling & Optimization of Dye Sensitized Solar Cells with Core/Shell Nanowire Array-Based Photoanodes

Justin J. Hill[1], Kirk J. Ziegler[1] and Nick Banks[2], (1)Chemical Engineering, University of Florida, Gainesville, FL, (2)University of Florida, Gainesville, FL

The benefits of utilizing nanowire-based photoanodes for dye sensitized solar cells is quickly becoming apparent. Nanowire structures allow for the presence of an interfacial electric field that is otherwise absent in nanoparticle-based dye sensitized solar cells (DSSCs). This electric field enhances charge transport by inducing potential gradient migration. This mechanism does not occur in nanoparticle-based devices, which are limited to charge transport by only diffusion. Additionally, the presence of an interfacial electric field is believed to reduce the rate of photogenerated electron back reaction with the electrolyte, enhancing photocurrent and allowing for differing electrolyte compositions that may increase photovoltage. Furthermore, the use of conductive core/semiconductive shell nanowire structures decouples the length a photogenerated charge must travel to be collected from the photoanode thickness (and thus overall photoactive surface area).

In this computational study, we investigate how the presence of an interfacial electric field affects interfacial charge transfer, bulk charge transport and charge carrier distributions. The numerical results indicate that electron loss reactions are 1000-fold less when an interfacial electric field is present. Furthermore, varying the special parameters of the photoanode to optimize performance indicates that DSSCs with photoanodes much thicker than 10 μm (the optimal thickness for nanoparticle-based DSSCs) can be used. This result indicates that efficiencies much larger than the current 12% achievement are attainable through conductive core/semiconductive shell nanowire arrays.

See more of this Session: Nanomaterials for Photovoltaics I
See more of this Group/Topical: Topical 5: Nanomaterials for Energy Applications

Using Supercritical Fluids for CdS and CdTeS Nanoparticles Decorated TiO_2 Nanostructures

Yaocihuatl Medina-Gonzalez,
*Nasrin Farhanghi and Paul A. Charpentier**

Department of Chemical and Biochemical Engineering, Faculty of Engineering, University of Western Ontario, London, Ontario, Canada N6A 5B9.

Phone: (519) 661 -3466. Fax: (519) 661-3498. E-mail: pcharpentier@eng.uwo.ca.

Abstract

An easy process was developed to synthesize TiO_2 nanowires decorated with CdS and CdTeS quantum dots (QDs) requiring no pretreatment of the TiO_2 nanowires prior to nanoparticle generation. CdS and CdTeS nanoparticles were firstly grown by an *in situ* colloidal method directly on the TiO_2 surface using both organic solvents and supercritical CO_2, hence not requiring subsequent functionalization of the QDs. The resulting nanostructure assembly and composition was confirmed by transmission electron microscopy (TEM), X-ray photoelectron spectroscopy (XPS) and Raman spectroscopy. Succesful decoration of TiO_2 nanowires by the QDs was observed by TEM, while XPS spectra provided clear evidence for the coexistence of CdS and CdTeS QDs and TiO_2 nanowires. The electronic structure of TiO_2 nanowires was preserved as indicated by Raman spectroscopy. This study shows that the generation of strongly anchored CdS and CdTeS QDs on TiO_2 nanowires surface without introduction of a linker molecule is possible. This work opens the possibility for the development of more efficient nanomaterials for electronic and photovoltaic devices and sensors where presence of a linker molecule is especially undesirable as its presence strongly decreases the electron injection efficiency. The method developed in this study represent a step forward in obtaining nanoparticles by greener processes like supercritical CO_2, and with a reduced quantity of process steps.

Keywords: decorated TiO_2, TiO_2 nanowires, CdS quantum dots, CdTeS quantum dots, assembly.

Introduction

Environmentally clean alternative energies that are carbon neutral are of tremendous current interest for a sustainable energy future. [1] In this context, solar energy is projected to play a prominent role as a primary energy source in the future. In the last decade the design of assemblies using semiconductor structures and metal nanoparticles has gained tremendous interest while considerable efforts have been undertaken to obtain high efficiency energy conversion devices. To enhance the efficiency, increasing the harvest of photons employing nanostructured semiconductor assemblies is required.[2] In this context, Titania (TiO_2) has been a well-studied semiconductor photoactive material with its major challenge being its large band-gap energy (3.0-3.2 eV) preventing its efficient visible light absorption.[3,4] Narrow band-gap inorganic semiconducting materials including CdS,[5-7] PbS,[8-10] Bi_2S_3,[8,11] CdSe,[12,13] InP[14] have been considered as promising sensitizers to enhance the utilization of sunlight for energy production in photovoltaic (PV) devices. In particular, cadmium calchogenide semiconducting nanocrystals (CdX; X=S, Se and Te) belonging to II-VI semiconductors are very attractive as sensitizers for TiO_2 due to their size-tunable optical properties.[15,16] Previous studies have shown that chemically and electrochemically deposited CdS, CdSe nanocrystallites on TiO_2 [13,17-20] and ZnO[21,22] surfaces are capable of injecting excited electrons that generate photocurrent under visible light radiation. In comparison, other quantum dots (QDs) with interesting emission properties in the near IR have not yet been as extensively examined. CdTeS nanoparticles are potentially superior to CdS for several applications including solar cells because of their additional near-IR activity.[23-25]

One-dimensional TiO_2 nanostructures (nanotubes, nanowires) as supports of light harvesting semiconducting nanocrystals have been shown to be more effective than TiO_2 nanoparticles in PV applications as their precisely oriented nature makes them excellent percolation pathways for vectorial charge transfer.[13,18,26] The surrounding "shell" of organic ligands that stabilizes the QDs dictates its surface chemistry. This shell plays a significant role in the solubility and the reactivity within a given environment. Ligand exchange is often required for the introduction of new surface functionality to QDs.[27-29] PbS and CdS QD's have been anchored to TiO_2 nanostructures in previous studies by using anchoring molecules including carboxylic acids and more precisely oleic and mercaptoalkanoic acids either by ligand exchange over the nanoparticles or by TiO_2 functionalization.[13,17,18,30,31] Considerable efforts have been made towards controlling the assembly of nanoparticles onto carbon nanotubes by techniques like laser pruning.[32] As well, CdS nanoparticles have been anchored to the TiO_2 nanotubes by sonoelectrochemical methods.[33] All these methods add an additional step in obtaining the nanoparticle decorated TiO_2, which leads to an expensive and difficult process to efficiently attach hydrophobic nanoparticles to the hydrophilic TiO_2 surface. In addition, the relative efficiencies of electron injection from photoexcited quantum dots into the molecularly linked TiO_2 nanoparticles depends on the chain length of the linker, i.e. as the chain length increases, the electron injection efficiency decreases dramatically[34], recent works have shown that the presence of ligands used to synthesize QDs also leads to poor conductivity in QDs films.[35]

Herein, we present a simple method to prepare TiO_2 nanowires decorated with CdS or CdTeS QDs without the need for a surface functionalization step for anchoring the QDs to the TiO_2 nanoparticles. This method allowed the assembly of strongly anchored CdS QDs onto the TiO_2 nanowires and was easily extended to CdTeS QDs. These assemblies are promising materials with applications in photovoltaic devices and sensors and the method developed represents a greener route to obtain these nanomaterials.

Experimental Section

Materials. Instrument grade CO_2 (99.5% purity) was purchased from Praxair, cadmium chloride (tech grade, ACS reagent), sodium diethyldithiocarbamate trihydrate (ACS reagent), Titanium isopropoxide (TIP) (97%), glacial acetic acid (99.99%), Tellurium granules (99.99%), trioctylamine (TOA) (99.0%) (Fluka), trioctylphosphine (TOP) (90%) (Fluka), anhydrous methanol (99.8%), toluene (99.9%) (CHROMASOLV Plus), lithium iodide (99.9%), iodine (99.8%), acetonitrile (anhydrous, 99.8%), sodium sulfide nonahydrate (99.99%), sulfur (99%), sodium hydroxide (99.9%) and Titanium (IV) oxide, anatase nanopowder <25nm (99.7%) were purchased from Sigma-Aldrich Canada. Anhydrous ethyl alcohol was purchased from Commercial Alcohols, Inc., Canada. 1,2-

Dimethyl-3-propylimidazolium iodide (DMPII) (98%) and 4-tert-butylpyridine (98%) were purchased from TCI America. All this products were used as received. ITO coated glass slides with a resistivity of 8-12 were purchased from Sigma-Aldrich and thoroughly washed with a soap solution, then with acetone under sonication. High purity conductive silver paint was obtained from SPI Supplies.

TiO_2 nanofibers synthesis. The schematic diagram of the experimental setup and details of the synthesis procedure for preparing TiO_2 nanofibers can be found in Sui et al. [36] In a typical experiment, 3.5 ml of TIP and 4 ml of acetic acid were quickly placed in a 10 ml view cell. The cell was closed and CO_2 was pumped into the view cell using a syringe pump (Isco 260D). Final temperature and pressure were maintained at 60°C and 5000 psig respectively. Stirring was allowed for 10 hrs. After 5 days of aging, washing with CO_2 was carried out at 0.2 ml/min for 24 hrs with the pressure then released and the solid recovered from the view cell.

CdS QDs synthesis. Preparation of the Single-Molecular Precursor Bis-(diethyldithyocarbamato) cadmium (II) ($Cd[S_2CN(C_2H_5)_2]_2$). The protocol for CdS QD's preparation has been previously described. [37] Stoichiometric amounts of a 0.1M aqueous solution of $NaS_2CN(C_2H_5)_2 \cdot 3H_2O$ were reacted with 0.1 M $CdCl_2$ solution to give a white precipitate that was further filtered under vacuum and washed by means of redispersion of the precipitate into distilled water followed by vacuum filtration. The product was subsequently dried by placing it in a vacuum oven at 40 °C overnight.

Preparation of CdS QD's. The method followed for CdS QDs preparation has been presented elsewhere.[38] In a typical experiment, a 25 mL sample of TOA and a magnetic stirring bar were placed in a 100 mL three-necked flask under nitrogen equipped with a reflux condenser, a thermometer, and a thermocouple for automatic temperature control. When the temperature was stable at 235 °C, a solution of 0.5 g of $Cd[S_2CN(C_2H_5)_2]_2$ in 9 mL of TOP was rapidly injected into the flask. 30 minutes after the addition of the cadmium diethyldithiocarbamate solution, the heater was removed to cool the reaction mixture. When the temperature dropped to approximately 75 °C, a large excess of methanol/ ethyl alcohol was added followed by separation of the quantum dots through centrifugation. The nanoparticles were washed with methanol, and then dispersed into toluene. The mixture of nanoparticles in toluene was further filtered under vacuum to remove any insoluble material.

Preparation of CdTeS QD's. The CdTeS QDs were synthesized using a modification of the procedure described before for CdS. A solution of 0.195g of Te in 2 ml of TOP was prepared. 3 minutes after the $Cd[S_2CN(C_2H_5)_2]_2$ adding to the three-neck flask, 0.3 ml of the Te/TOP solution was added. After 3 minutes, other 0.3 ml were again added. After 3 minutes, 0.9 ml of the Te/TOP solution were added drop by drop. The rest of the synthesis was similar to CdS nanoparticles preparation. The total reaction time was 30 minutes.

TiO_2 nanofibers decorated with CdS and CdTeS QD's. 0.3 g of TiO_2 nanofibers prepared by using supercritical CO_2 as solvent, were calcined at 400°C and added to a three-neck flask containing 25 ml of TOA and sonicated for 2h. CdS or CdTeS QDs were prepared as described in 2.2. The solids obtained were thoroughly washed with toluene under sonication for 30 minutes, filtered to remove the nanoparticles weakly bonded to the TiO_2 nanowires and dried under vacuum overnight at 40°C.

Characterization. Photoluminescence spectra (PL) were recorded using a Quantamaster 50 spectrofluorometer equipped with a Xenon short arc lamp. Raman spectra were effectuated using a

Kaiser Optical Systems RXN1-785 instrument with an excitation wavelength of 785 nm. Scanning Electron Microscopy (SEM) micrographs were obtained using a JEOL 2010F, while Transmission Electron Microscopy (TEM) images were obtained with a Philips CM-10 Transmission Electron Microscope with digital camera. Samples for SEM imaging were prepared by applying the powder directly to a carbon adhesive tape without metallization. High Resolution Transmission Electron Microscopy (HRTEM) images and X-Ray Energy Dispersive Analysis (EDS) were obtained with a JEOL JEM 2010F Field Emission Electron Microscope equipped with a Gatan imaging filter. TEM, HRTEM and EDS images were aquired after dispersion of the nanocomposites in toluene by sonication. XRD spectra were collected on a Rigaku-Miniflex bench-top X-ray powder diffractometer (Carlsbad, CA). Using CuKα (λ for Kα = 1.54059Å) radiation obtained at 30 kV and 15 mA. XPS analyses were carried out with a Kratos Axis Ultra spectrometer using a monochromatic Al K(alpha) source (15mA, 14kV). Sonication was carried out by VWR aqua sonic (model 50 D).

Results and Discussion

Synthesis. As shown schematically by Figure 1a, sonication of the TiO_2 nanowires in trioctylamine (TOA) solution prior to *in situ* growing of the CdS nanoparticles in this solvent-stabilizer allowed the direct obtainment of CdS nanoparticle decorated TiO_2 nanowires. Typically, TiO_2 nanowires are completely hydrophilic, unable to be dispersed in toluene even after sonication. A clear changing in solubility behavior was observed after assembly of CdS and CdTeS QDs onto the TiO_2 nanowires. After assembly, they showed a hydrophobic nature, being well dispersed in toluene using the sonication method (Figure 1b).

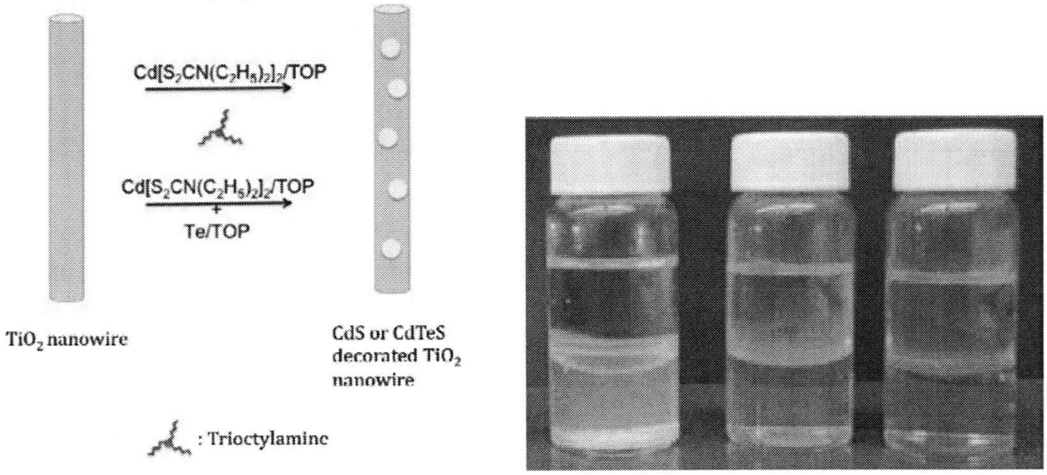

A B

Figure 1. A) Schematic of CdS and CdTeS QD's /TiO_2 nanowire assembly; and B) Photos of (from left to right) TiO_2 nanowires, CdS/TiO_2 and CdTeS/TiO_2 in a mixture toluene (top)-water (bottom) after 3 days (dispersion prepared by sonication).

Morphology. Figure 2a shows the TEM images of the TiO_2 nanowires obtained by the sol-gel method in supercritical CO_2, while Figure 2b and 2c shows the CdS and CdTeS QDs anchored to the TiO_2 nanowire surfaces, respectively. It is apparent that the nanoparticles are directly attached to the titania surface as further discussed below. Figure 2d provides a HRTEM image of the CdS/TiO_2 assembly in which the CdS nanocrystals have grown directly on the TiO_2 wires. CdTeS/TiO_2 HRTEM-EDS elemental mapping is shown in Figure 2e in which the Ti, O, Cd, Te and S elemental distributions can be observed (the elemental distributions for CdS/TiO_2 are provided in SI Figure 2). In both cases, excellent distribution of Ti and O is evident in the rods, while Cd, S and Cd, Te, and S elements were detected in the QDs of the CdS/TiO_2 and $CdTeS/TiO_2$ assemblies, respectively.

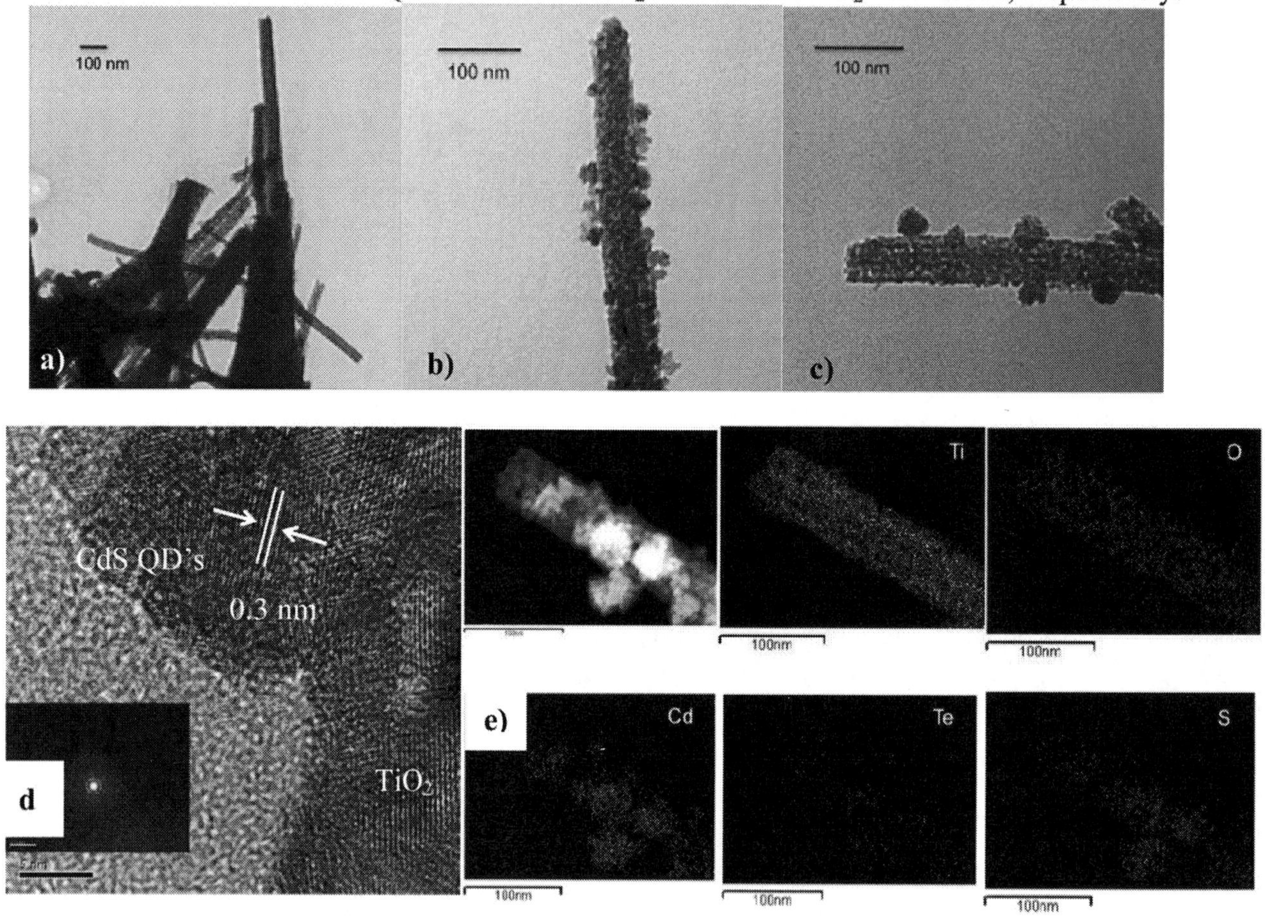

Figure 2. TEM images of a) TiO_2 nanowires, b) CdS/TiO_2 nanowires and c) $CdTeS/TiO_2$ nanowires. d) HRTEM image of CdS/TiO_2, e) HRTEM-EDS elemental mapping of $CdTeS/TiO_2$ assembly.

Typical XRD patterns of the CdS and CdTeS QDs are shown in Figure 3a. These diffraction patterns are consistent with the lattice parameters for cubic CdS and CdTeS QDs i.e. the interlayer distances measured were 0.3 nm for CdS and 0.355 nm for CdTeS QDs. A strong peak is observed at $2\theta=26.6°$ as a result of the diffraction of the (111) plane of cubic CdS, while two other peaks can be observed at 43.55° and 48° and are indexed to the (220) and (311) planes of cubic CdS respectively[39,40] while CdTeS QDs showed similar patterns. These XRD peaks are diffused and broadened as a result of the small size of the crystallites. These results confirm cubic CdS and CdTeS phases on the surface of the TiO_2 nanorods, as can be observed on the XRD patterns of CdS/TiO_2 and $CdTeS/TiO_2$ assemblies shown in Figure 3b.

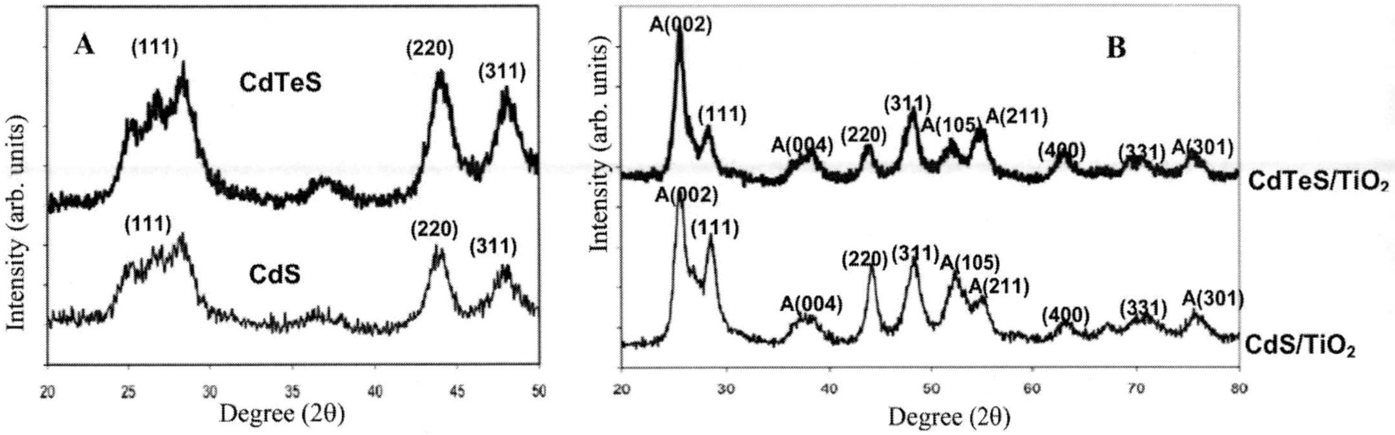

Figure 3. XRD pattern of A) CdS and CdTeS QD's and B) CdS and CdTeS/TiO_2 assemblies. The peaks corresponding to cubic CdS and CdTeS and to anatase phase of TiO_2(A).

CdTeS showed a high PL peak in the near IR region (650-800 nm) (Figure 4a) as a result of Te inclusion in good agreement with previous studies[23,24] while CdS showed a peak with a maximum around 540nm as expected from previous results.[37] UV-vis spectra of both the CdS and CdTeS QDs are shown in Figure 4b. As expected, inclusion of Te increased the light absorption compared to CdS.[24,41]

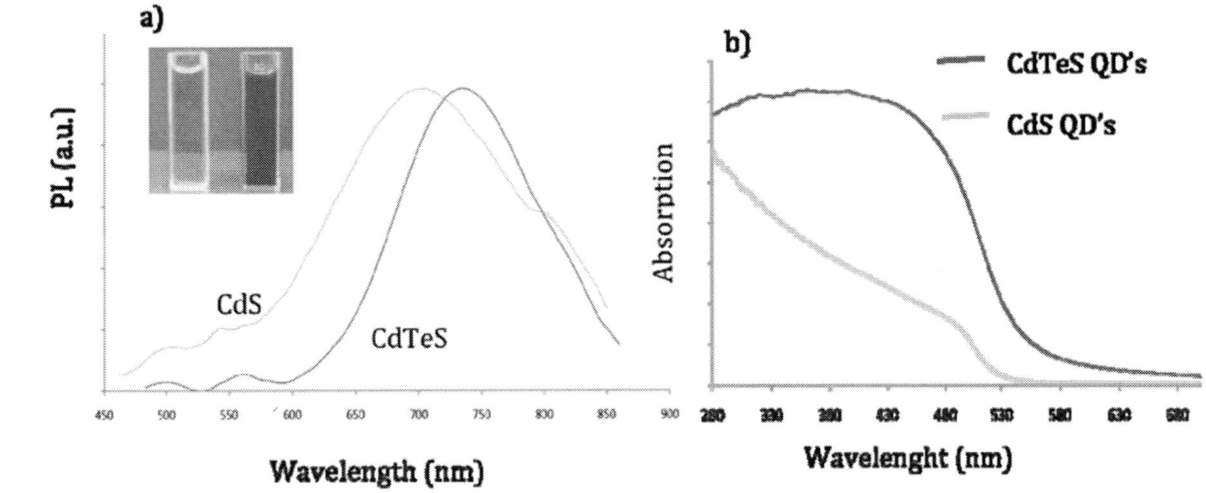

Figure 4. a) Photoluminescence spectra of CdS and CdTeS QDs (excitation wavelength: 430 nm). Curves have been shifted on purpose for clarity. Inset shows CdS (yellow) and CdTeS (orange) QDs in toluene, b) UV-vis spectra of CdS and CdTeS and QD's.

Raman spectra for the TiO_2 nanowires and QDs assemblies are shown in Figure 5. The three spectra show the characteristic peaks of TiO_2 that are similar to that of titanate.[18] The similarity in the peak intensities indicates a preserved structure of the TiO_2 nanowires meaning that the CdS or CdTeS QD attachment to TiO_2 is non-destructive. Two signals corresponding to the longitudinal optical phonon mode (1-LO) and its overtone (2-LO) of the CdS nanoparticles at *ca.* 300cm^{-1} and 580cm^{-1} were found.[42,43] The signals found at *ca.* 300 cm^{-1} and between 160-180 nm^{-1} in CdTeS decorated TiO_2 spectrum were attributed to CdS-like 1-LO phonons and to CdTe like 1-LO phonons respectively. [44,45] The signals from the QDs are weak because the most abundant compound in the assemblies is TiO_2.

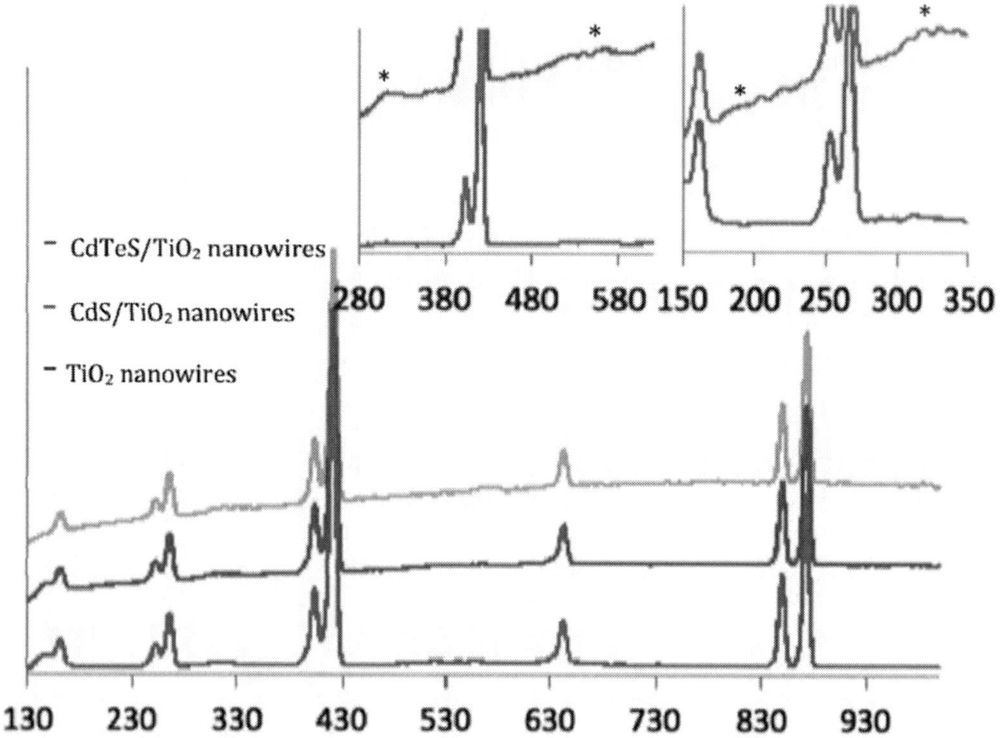

Figure 5. Raman spectra of TiO_2 nanowires before and after anchoring of CdS and CdTeS QDs. Insets show enlarged view. Spectra have been shifted for clarity. The symbol (*) shows the signals attributed to CdS and CdTeS QDs phonons.

XPS spectra provided more detailed structural information on the QD/TiO_2 assemblies. The spin-orbit components ($2p_{3/2}$ and $2p_{1/2}$ and) of the Ti2p are well deconvoluted into two peaks at 458.5 and 464.2 eV for TiO_2-CdS and at 458.14 eV and 463.86 eV for TiO_2/CdTeS, corresponding to Ti^{4+} in a tetragonal structure.[46] The O1s showed a peak that was deconvoluted into two peaks at 530.98 eV and 529.76 eV for TiO_2-CdS and at 530.3 eV and 529.38 for TiO_2-CdTeS that can be attributed to Ti_2O_3 and TiO_2. This Ti_2O_3 can be formed by reduction reactions from organic radicals during heat treatment of TiO_2.[47] The $Cd3d_{5/2}$ and $Cd3d_{3/2}$ were found at 411 and 404.8 eV respectively for TiO_2-CdS and at 411.1 eV and 404.41 eV for TiO_2-CdTeS, which are attributed to the Cd^{2+} of CdS. The S2p peak was found at 160.1 eV, attributable to the S^{2-} of CdS. For TiO_2-CdTeS, S2p1/2 and S2p3/2 were found in the range 161.95-163 eV and 160.77-161.82 respectively. There can exist some contribution to these peaks coming from S2p bound to TiO_2 in the form of S^{2-} [48] which would help explain the attachement of CdS and CdTeS nanoparticules to the TiO_2 nanowires through an electron donor-acceptor mechanism. In addition, short range forces associated to strong van der Waals forces are present at this nanoscale level between nanowires and QDs,[49] given the dipolar moments present in semiconductors quantum dots[50,51] accentuated when electron acceptor atoms such as S are present, together with permanent dipole moments existing on the TiO_2 surface originated by undercoordinated Ti atoms.[52] These dipole-dipole interactions would be strong driving forces for direct assembly in our systems. Te3d5/2 and Te3d3/2 appeared at 571.6 and 582eV respectively while the splitting observed in the Te3d peak is due to the presence of Te-Cd bonding in CdTeS.[53] After calculations form XPS, the concentrations of atoms in the surface were $Cd_{0.92}S_{0.08}$ and $Cd_{0.8}Te_{0.07}S_{0.1}$ for CdS and CdTeS respectively.

Further research is needed in order to elucidate if TOA molecules are incorporated as linker between the QDs and TiO_2 nanowires or their role is merely as solvent and stabilizant during QDs formation.

The energy diagram of the prepared assemblies is shown in Figure 7. In this figure the bulk bandgaps and conduction band (Ec) position for CdS and TiO_2 were taken from the literature,[17] the bulk band gap for CdTeS was taken from Muzafarova et al.,[54] while the conduction and valence band (Ev) positions were calculated from this study. CdS decorated TiO_2 and CdTeS decorated TiO_2 assemblies are type II heterostructures, as both the valence and the conduction bands in TiO_2 are lower than those in CdS and CdTeS quantum dots. In this heterostructure, the band gap offset can be tuned by increasing the Te concentration [55] which makes us belief these hybrid materials can be successfully applied to photovoltaic devices and electronic sensors.

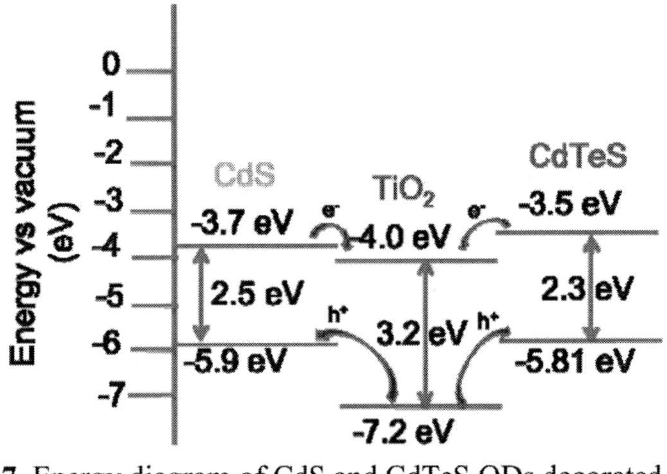

Figure 7. Energy diagram of CdS and CdTeS QDs decorated TiO_2.

Conclusions

A simple method was developed for preparing CdS and CdTeS quantum dots anchored onto TiO_2 nanowires without a pretreatment step or ligand exchange. The method developed allows strong anchoring of CdS and CdTeS on TiO_2 nanowires surface without the need of a linker molecule whose presence is known to decrease the electron injection efficiency when this type of assemblies are used in photovoltaic devices as photoactive layer. Our method makes hybrid materials solution-processable. This assembly method can be easily extended to produce various nanoparticle-nanowire materials. It should be noted that it would be very interesting to investigate the photovoltaic properties of this hybrid materials. Finally, this work provides us the opportunity for fundamental nanoscience research, as well to open the possibility for the development of more efficient nanomaterials for electronic and photovoltaic devices and sensors by greener routes.

References

(1) Hemminger, J. *Directing Matter and Energy: Five Challenges for Science and the Imagination*; A repport from the Basic Energy Sciences Advisory Comitee, USA Department of Energy. Report available online at http://www.sc.doe.gov/bes/reports/list.html. , 2007.

(2) Kamat, P. V. *The Journal of Physical Chemistry C* **2008**, *112*, 18737.

(3) Linsebigler, A. L. L., Guangquan; Yates, John T. , Jr. . *Chem. Rev.* **1995**, *95*, 735.

(4) Tracy L.Thompson and John T. Yates, J. *Chem. Rev.* **2006**, *106*, 4428.

(5) Lübke, H. G. M. *Journal of Electroanalytical Chemistry* **1986**, *204*, 225.

(6) R. Vogel, K. P., H. Weller. *Chem. Phys. Letters* **1990**, *174*, 241.

(7) Sakata, S. K. A. K. T. *Chem. Phys. Letters* **1993**, *206*, 166.

(8) Vogel, R., Hoyer, P., Weller, H. . *J. Phys. Chem.* **1994**, *98*, 3183.

(9) Plass, R. P., Serge; Krueger, Jessica; Graetzel, Michael; Bach, Udo. *J. Phys. Chem. B* **2002**, *106*, 7578.

(10) Lee, H. L., Henry C.; Moon, Soo-Jin; Chen, Peter; Ito, Seigo; Haque, Saif A.; Torres, Tomas; Nueesch, Frank; Geiger, Thomas; Zakeeruddin, Shaik M.; Graetzel, Michael; Nazeeruddin, Md. Khaja. *Adv. Funct. Mat.* **2009**, *19*, 2735.

(11) Peter, L. M. W., K. G. Upul; Riley, D. Jason; Waggett, Jonathan P. . *J. Phys. Chem. B* **2003**, *107*, 8378.

(12) Liu, D. K., Prashant V. . *J. Phys. Chem.* **1993**, *97*, 10769.

(13) Kongkanand, A.; Tvrdy, K.; Takechi, K.; Kuno, M.; Kamat, P. V. *Journal of the American Chemical Society* **2008**, *130*, 4007.

(14) Zaban, A. M., O. I.; Gregg, B. A.; Nozik, A. J. . *Langmuir* **1998**, *14*, 3153.

(15) Peng, Z. A.; Peng, X. *Journal of the American Chemical Society* **2000**, *123*, 183.

(16) Robel, I.; Kuno, M.; Kamat, P. V. *Journal of the American Chemical Society* **2007**, *129*, 4136.

(17) Das, K.; De, S. K. *The Journal of Physical Chemistry C* **2009**, *113*, 3494.

(18) Feng, M.; Zhan, H.; Miao, L. *ACS Applied Materials \& Interfaces* **2010**, *2*, 1129.

(19) Nasr, C.; Hotchandani, S.; Kim, W. Y.; Schmehl, R. H.; Kamat, P. V. *The Journal of Physical Chemistry B* **1997**, *101*, 7480.

(20) Tachibana, Y.; Akiyama, H. Y.; Ohtsuka, Y.; Torimoto, T.; Kuwabata, S. *Chemistry Letters* **2007**, *36*, 88.

(21) Hotchandani, S. K., Prashant V. . *Chemical Physics Letters* **1992**, *191*, 320.

(22) Hotchandani, S. K., Prashant V. . *J. Phys. Chm.* **1992**, *96*, 6834.

(23) Zhang, S. W., Wenzhi; Song, Weisi; Wang, Yinghui; Peng, Yajing; Liu, Yuqiang; Yang, Yanqiang. *Optik* **2010**, *121*, 312.

(24) Yudong Liu, V. C. S. R., R. E C. Royas C. H. Brito Cruz C. L. Cesar H. L. Fragnito L. C. Barbosa; Alves, O. L. *Journal of Materials Science Letters* **1996**, *15*, 980.

(25) Jiaxiang Zhang, X. H., Qihuang Gong. *Physics Letters A* **2009**, *373*, 4475.

(26) Baker, D. R.; Kamat, P. V. *Advanced Functional Materials* **2009**, *19*, 805.

(27) Chan Warren, C. W., Nie, Shuming. *Science* **1998**, *281*, 2016.

(28) Mattoussi, H.; Mauro, J. M.; Goldman, E. R.; Anderson, G. P.; Sundar, V. C.; Mikulec, F. V.; Bawendi, M. G. *Journal of the American Chemical Society* **2000**, *122*, 12142.

(29) Wang, Y. A.; Li, J. J.; Chen, H.; Peng, X. *Journal of the American Chemical Society* **2002**, *124*, 2293.

(30) Wang, D.; Zhao, H.; Wu, N.; El Khakani, M. A.; Ma, D. *The Journal of Physical Chemistry Letters* **2010**, *1*, 1030.

(31) Sun, W.-T.; Yu, Y.; Pan, H.-Y.; Gao, X.-F.; Chen, Q.; Peng, L.-M. *Journal of the American Chemical Society* **2008**, *130*, 1124.

(32) Lim, X.; Zhu, Y.; Cheong, F. C.; Hanafiah, N. M.; Valiyaveettil, S.; Sow, C.-H. *ACS Nano* **2008**, *2*, 1389.

(33) ChengLin Wang, L. S., Hong Yun Jing Li YueKun Lai; Lin, C. *Nanotechnology* **2009**, *20*, 295601.

(34) Dibbell, R. S.; Watson, D. F. *The Journal of Physical Chemistry C* **2009**, *113*, 3139.

(35) Williams, K. J.; Tisdale, W. A.; Leschkies, K. S.; Haugstad, G.; Norris, D. J.; Aydil, E. S.; Zhu, X. Y. *ACS Nano* **2009**, *3*, 1532.

(36) Sui, R.; Rizkalla, A. S.; Charpentier, P. A. *The Journal of Physical Chemistry B* **2006**, *110*, 16212.

(37) Xu, W. Z.; Charpentier, P. A. *The Journal of Physical Chemistry C* **2009**, *113*, 6859.

(38) Murray, C. B.; Norris, D. J.; Bawendi, M. G. *Journal of the American Chemical Society* **1993**, *115*, 8706.

(39) Feng, M.; Sun, R.; Zhan, H.; Chen, Y. *Carbon* **2010**, *48*, 1177.

(40) Ji, K. H.; Jang, D. M.; Cho, Y. J.; Myung, Y.; Kim, H. S.; Kim, Y.; Park, J. *The Journal of Physical Chemistry C* **2009**, *113*, 19966.

(41) Jin Ho Bang, W. H. S., and Kenneth S. Suslick. *Chem. Mater.* **2008**, *20*, 4033.

(42) V. Sivasubramanian, A. K. A., M. Premila C.S. Sundar; Sastry, V. S. *Physica E* **2006**, *31*, 93.

(43) Li, X.; Jia, Y.; Cao, A. *ACS Nano* **2009**, *4*, 506.

(44) Fischer, A.; Feng, Z. *Applied Physics Letters* **1997**, *70*, 3239.

(45) Fischer, A.; Anthony, L.; Compaan, A. D. *Applied Physics Letters* **1998**, *72*, 2559.

(46) Juhong Zhu, D. Y., Jiaqing Geng Daimei Chen; Jiang, Z. *J. Nanopart. Research.* **2008**, *10*, 729.

(47) Jimmy C. Yu, J. Y., Hung Yuk Tanga; Zhang, L. *J. Mater. Chem.* **2002**, *12*, 81.

(48) E.L.D. Hebenstreit, W. H.; Diebold, U. *Surface Science* **2001**, *470*, 347.

(49) Rance, G. A.; Marsh, D. H.; Bourne, S. J.; Reade, T. J.; Khlobystov, A. N. *ACS Nano* **2010**, doi: 10.1021/nn101287u.

(50) Tang, Z.; Kotov, N. A.; Giersig, M. *Science* **2002**, *297*, 237.

(51) Talapin, D. V.; Shevchenko, E. V.; Murray, C. B.; Titov, A. V.; Kral, P. *Nano Letters* **2007**, *7*, 1213.

(52) Alimohammadi, M.; Fichthorn, K. A. *Nano Letters* **2009**, *9*, 4198.

(53) K.R. Murali, P. T., C. Kannan; Sengodan, V. *Solar Energy* **2009**, *83*, 14.

(54) S. A. Muzafarova, B. U. A., Sh. A. Mirsagatov, K. Durshimbetov, and Zh. Zhanabergenov. *Semiconductors* **2008**, *42*, 1377.

(55) O□zden Akinci, H. H. G. r., Hilmi U□nlu□. *Thin Solid Films* **2009**, *517*, 2431.

(413f) Evidence for High-Efficiency Exciton Dissociation at Poly(3-hexylthiophene)/Single-Walled Carbon Nanotube Interfaces in Planar Nano-Heterojunction Photovoltaics

Moon-Ho Ham[1], Geraldine LC Paulus[1], Chang Young Lee[1], Changsik Song[1], Kourosh Kalantar-zadeh[2], Won Joon Choi[1], Jae-Hee Han[1] and Michael S. Strano[1], (1)Department of Chemical Engineering, Massachusetts Institute of Technology, Cambridge, MA, (2)School of Electrical and Computer Engineering, RMIT University, Melbourne, Australia

There is significant interest in combining carbon nanotubes with semiconducting polymers for photovoltaic applications due to potential advantages from smaller exciton transport lengths and enhanced charge separation. However, to date, bulk heterojunction (BHJ) devices have demonstrated relatively poor efficiencies, and little is understood about the polymer/nanotube junction. To investigate this interface, we fabricate a planar nano-heterojunction comprised of well-isolated millimeter-long single-walled carbon nanotubes underneath a poly(3-hexylthiophene) (P3HT) layer. The resulting junctions display photovoltaic efficiencies per nanotube ranging from 3% to 3.82%, which exceeds those of polymer/nanotube BHJ by a factor of 50–100. The increase is attributed to the absence of aggregate formation in this planar device geometry. It is shown that the polymer/nanotube interface itself is responsible for exciton dissociation. Typical open-circuit voltages are near 0.5 V with fill factors of 0.25–0.3, which are largely invariant with the number of nanotubes per device and P3HT thickness. A maximum efficiency is obtained for a 60 nm-thick P3HT layer, which is predicted by a Monte Carlo simulation that takes into account exciton generation, transport, recombination and dissociation. This platform is promising for further understanding the potential role of polymer/nanotube interfaces for photovoltaic applications.

See more of this Session: Nanomaterials for Photovoltaics I
See more of this Group/Topical: Topical 5: Nanomaterials for Energy Applications

(479a) Performance Enhancement of Hybrid Solar Cells through Chemical Vapor Annealing

Yue Wu and Genqiang Zhang, School of Chemical Engineering, Purdue University, West Lafayette, IN

Semiconductor nanostructure-based solar cells, such as dye-sensitized solar cells, all-inorganic nanostructure solar cells, and hybrid inorganic nanostructure-conductive polymer composite solar cells, are promising candidates for future-generation photovoltaic devices due to their advantages in low temperature processing, scalability, and low cost. Among these different types of nanostructurebased solar cells, hybrid solar cells, which originated from the concept of organic polymer bulk heterojunction solar cells, possess a unique architecture of a solid-state blend of interpenetrating and percolating inorganic nanostructures and organic polymers. In principle, hybrid solar cells should demonstrate a better performance when compared with organic polymer bulk heterojunction solar cells due to the higher absorption coefficient of inorganic semiconductor nanomaterials. However, so far there has been no report of higher performance observed in hybrid solar cells than organic bulk heterojuction solar cells in the real experiment, which can be mainly attributed to the complicated surface/interface of nanomaterials, for example, the existence of non-conductive surfactant molecules with long alkyl chain such as phosphonic acids on the surface of nanocrystals. Notably, in all-inorganic nanocrystal solar cell system, this problem could be solved through surface ligand exchange reactions by replacing of long alkyl chain surfactants with short chain molecules, for example, chloride, hydrazine, amine, thiol, etc. After the ligand exchange reactions, the inter-particle distance in the nanocrystal films is getting smaller, thus facilitate the electron/hole transport. It would be ideal that a similar approach could be developed during the solution process of nanocrystal/conductive polymer hybrid solar fabrication. However, practically it still remains a challenge because once the long alky chain ligands on the nanocrystal surface are replaced with small molecules, the nanocrystals tend to aggregate and precipitate out from the organic solvent, thus cannot be mixed uniformly with conductive polymer any more for solution process. On the other hand, solvent vapor annealing process has been demonstrated as an effective approach to improve crystallinity of organic semiconductor thin films and enhance their electrical properties. Herein, we report the first demonstration of how a straightforward, one-step, chemical vapor annealing process involving surface ligand exchange reaction can dramatically improve the power conversion efficiency of a hybrid solar cell composed of cadmium selenide (CdSe) nanorods, as an electron acceptor and a conducting polymer, poly(3-hexylthiophene) (P3HT), as an electron donor, without the sacrifice of solution processability of nanocrystals in organic solution and polymer blend.

See more of this Session: Nanomaterials for Photovoltaics II
See more of this Group/Topical: Topical 5: Nanomaterials for Energy Applications

(479b) Exploiting Equilibrium Properties for Assembly of Core/Shell-Like Compound Semiconductor Nanocrystals

Sumeet C. Pandey, T. J. Mountziaris and Dimitrios Maroudas, Department of Chemical Engineering, University of Massachusetts, Amherst, MA

Typical synthesis of core/shell nanostructures involves a rigorous two-step protocol. It is common to synthesize the semiconducting core first, followed by the overgrowth of a thin shell of a different band-gap material. The few attempted single-step and one-pot synthesis approaches have been limited in their use and specific to the type of precursors involved. The core/shell nanocrystals prepared by over-coating methods enclose interfaces of lattice-mismatched core and shell regions that lead to undesirable misfit strain and defect-induced states in the band gap. Motivated by the need for single-step synthesis routes to design minimally strained core/shell-like nanostructures, we study the equilibrium properties of a prototypical ternary compound semiconductor nanocrystal system, emphasizing on equilibrium surface segregation, as a potential means to self-assemble core/shell-like semiconductor quantum dots.
In this presentation, we report results of equilibrium surface segregation in the prototypical system, namely, $ZnSe_{1-x}Te_x$ nanocrystals based on a computational analysis of coupled compositional, structural, and volume relaxation of the nanocrystals. The analysis employs Monte Carlo and conjugate-gradient methods according to a description of interatomic interactions that has been parameterized based on first-principles density functional theory (DFT) calculations of surface segregation energies. We have determined the equilibrium concentration distribution as a function of nanocrystal size and composition for nanocrystal morphologies that include faceted equilibrium crystal shapes and spherical nanocrystals. The results identify the nanoparticle size and composition ranges that allow for self-assembly of core/shell-like nanocrystal structures that are characterized by Te-deficient core and Te-rich compositionally graded shell-like regions. In addition, core/shell ZnSe/ZnTe nanocrystals and the equilibrium configurations generated from the equilibrium segregation analysis in $ZnSe_{1-x}Te_x$ nanocrystals are compared and contrasted within the DFT framework. Results for core/shell-like structure formation due to equilibrium surface segregation also are presented for ternary compound semiconductor nanocrystals other than the prototypical system, including III-V compounds such as $In_xGa_{1-x}As$.

See more of this Session: Nanomaterials for Photovoltaics II
See more of this Group/Topical: Topical 5: Nanomaterials for Energy Applications

(479c) Solution-Grown CuInSe2 Nanowires for Low-Cost Photovoltaics

Chet Steinhagen, Vahid Akhavan, Brian Goodfellow, Matthew G. Panthani, Justin Harris, Vincent C. Holmberg and Brian A. Korgel, Chemical Engineering, University of Texas at Austin, Austin, TX

A semiconductor media that could be printed under ambient conditions by a roll-to-roll process on plastic substrates to fabricate photovoltaic devices (PVs) would greatly reduce the cost of solar cell manufacturing. One approach is to create "paints" or "inks" of nanocrystals that can be deposited by solution deposition processes like casting, dip-coating or spraying. Recently, our group demonstrated that copper indium diselenide (CIS) nanocrystals can be synthesized and dispersed in organic solvents and then used to fabricate functioning PVs without the need for high temperature annealing or vacuum processing of the absorber layer. Unfortunately, the power conversion efficiency of these kinds of devices is still too low for commercialization, largely because of the losses in the film due to the high density of interfaces between neighboring nanocrystals. One approach to eliminating these interfaces, while retaining the printability of the media is to synthesize nanorods or nanowires that can be dispersed and printed. Here, we report the synthesis of CIS nanowires by solution-liquid-solid (SLS) and supercritical fluid-liquid-solid (SFLS) growth. Nanocrystals of bismuth are used as crystallization seeds of the nanowires in either a hot high-boiling solvent, or in an organic solvent under pressure at temperatures well above the critical point of the solvent. Significant quantities of nanowires can be synthesized using these approaches. Prototype PVs have been fabricated by spray-deposition of the nanowires. We also demonstrate the formation of a nanowire paper, or non-woven fabric, with the potential to serve as a new type of free-standing PV material.

See more of this Session: Nanomaterials for Photovoltaics II
See more of this Group/Topical: Topical 5: Nanomaterials for Energy Applications

(479d) Non-Thermal Plasmas and Semiconductor Nanocrystals

Lorenzo Mangolini and Uwe Kortshagen, Mechanical Engineering, University of Minnesota, Minneapolis, MN

Many of the processes of technological interest involving the use of plasmas are negatively affected by the nucleation and growth of particles in the gas-phase. Much of the research in the area of particles and plasmas, as a consequence, has focused on the elimination of particles from these processes. A radically new approach will be presented, in which nanoparticles are deliberately synthesized in a plasma and studied for a class of novel technological applications. In particular, this work focuses on the synthesis of silicon quantum dots, small silicon nanocrystals whose electronic and optical properties strongly deviate from those of the bulk material. A continuous flow non-thermal plasma reactor will be described in detail. In non-thermal plasmas, the gas temperature remains close to room temperature while the presence of energetic free-electrons triggers the nucleation and growth of nanocrystals. Moreover, the electrostatic charging of the particles slows down agglomeration, making this process ideal for the synthesis of small particles. For the first time, we have studied the nanoparticle energy balance in the discharge and shown that there is a significant non-equilibrium between the particle temperature and the background gas temperature, likely leading to the formation of high quality nanocrystals. This work also addresses the passivation and functionalization of the nanoparticles, with the goal of (1) maximizing the fluorescence efficiency of the crystals and (2) providing a good dispersion of particles in the desired solvent, enabling the formulation of inks for various applications, such as printed electronics or roll-to-roll fabrication of solar cells. In particular, I intend to highlight the importance of these novel nanomaterials for energy-related applications, such as photovoltaics.

See more of this Session: Nanomaterials for Photovoltaics II
See more of this Group/Topical: Topical 5: Nanomaterials for Energy Applications

(479e) Photocatalytic Activity of TiO2 and TiO2-xCxNy Thin Films From Polymer Assisted Deposition

Stacy Baber, Qianglu Lin, Venkata Daram, David Rockstraw, Shuguang Deng and Hongmei Luo, Chemical Engineering, New Mexico State University, Las Cruces, NM

Since the 1972 discovery of Fujishima and Honda detecting the occurrence of water splitting on TiO2 electrodes, titanium dioxide's use as a photocatalyst has provoked a broad range of research studies and applications. Applications like air and water-purifying, deodorizing, anti-bacterial use and self-cleaning are now being realized using. Pure TiO2's band gap is large (3.2 eV), causing its photocatalytic property to be active in the ultraviolet region which is <10% of the overall solar intensity. In this presentation, we discuss the synthesis and application of TiO2-based photocatalyst with enhanced visible light response. TiO2 and C-doped TiO2 films were grown on ITO glass by a chemical solution technique, polymer-assisted deposition (PAD) by heating the Ti-polymer precursor films at temperatures of 300 to 550 °C. Structure analyses by SEM, XRD, as well as UV-Vis spectra are reported. At 500 °C, the band gap was found to be ~3.05 eV and followed a linear relationship with increasing temperature as more of the polymer was burned off. TiO2 annealed at 350 °C had remaining polymer in the film, resulting in a lower band gap energy of ~1.95 eV. At temperatures lower than 400 °C, TiO2-xCx exists with a reduced band gap and should exhibit visible light photocatalytic properties. However, crystallinity of the C-doped TiO2 films is reduced to the point of films being amorphous, and results in the films showing no visible-light activity. This result is unique as it shows that one can reduced the band gap energy, but it does not guarantee enhanced photocatalytic efficiency.

See more of this Session: Nanomaterials for Photovoltaics II
See more of this Group/Topical: Topical 5: Nanomaterials for Energy Applications

(479f) Ordered TiO2 Nanotube Arrays On Transparent Conducting Oxide for Efficient Dye-Sensitized Solar Cells

Chengkun Xu, Chemical Engineering, University of Pittsburgh, Pittsburgh, PA and Di Gao, Chemical and Petroleum Engineering, University of Pittsburgh, Pittsburgh, PA

Synthesizing highly ordered 1-D TiO2 nanostructures directly on transparent conductive oxides (TCO), which is of key interest for fabricating front-side illuminating dye-sensitized solar cells (DSCs), has been one of the most challenging nanofabrication tasks. Here, we present a process for synthesizing vertically aligned TiO2 nanotube arrays on TCO through a liquid-phase conversion process using ZnO nanowire arrays as a template. The resulting TiO2 nanotube arrays can be easily integrated with the fabrication process of DSCs and lead to significantly improved photovoltaic performance compared to ZnO nanowire-based DSCs. It is also observed that the lifetime of photogenerated electrons in TiO2 nanotubes is more than an order of magnitude larger than that in sintered TiO2 nanoparticles. This result opens up numerous opportunities to further improve DSCs, for example, by employing solid-state electrolytes and redox mediators with faster kinetics, which otherwise would be challenging or impossible to implement by the use of nanoparticlebased DSCs.

See more of this Session: Nanomaterials for Photovoltaics II
See more of this Group/Topical: Topical 5: Nanomaterials for Energy Applications

(546a) $Cu(In,Ga)Se_2$ (CIGS) and $CuIn(SeS)_2$ (CISS) Nanocrystals and Their Use as Printable Photovoltaic Media and Medical Imaging Contrast Agents

Dariya K. Reid, Matthew G. Panthani, Mike Rasch, Vahid A. Akhavan, Brian W. Goodfellow and Brain A. Korgel, Chemical Engineering, University of Texas at Austin, Austin, TX

We have developed the colloidal synthesis of I-III-VI_2 semiconductor nanocrystals of $Cu(In,Ga)Se_2$ (CIGS) and $CuIn(SeS)_2$ (CISS). These nanocrystals can be spray-deposited into photovoltaic devices (PVs) with power conversion efficiencies of up to 3% without the use of high temperature annealing. Mechanically flexible PVs on plastic substrates can also be made using these materials. The gallium content can be varied systematically in the nanocrystals as well to tune the bandgap energy. By creating a library of CIGS nanocrystal inks with varying Ga composition, PVs with an absorber layer with a Ga composition gradient can be fabricated, yielding devices with graded bandgaps. The stability of the nanocrystal dispersions is important for creating uniform films of nanocrystals and functioning PVs. Typically, oleylamine is used as a capping ligand, but nanocrystals of relatively large diameter (> 15 nm) are usually obtained. By manipulating the ligand chemistry, nanocrystals with much smaller diameter, below the size needed for quantum confinement, can be synthesized. These CISS nanocrystals emit light in the red/near-IR region of the spectrum and are being examined as luminescent contrast agents for medical imaging.

See more of this Session: Nanomaterials for Photovoltaics III
See more of this Group/Topical: Topical 5: Nanomaterials for Energy Applications

(546b) Ordered Nanoarchitectures Improve Plasmon Enhancement near Semiconductor Bandgaps in the near-IR

D. Keith Roper, University of Arkansas, Fayetteville, AR, Wonmi Ahn, Materials Science and Engineering, University of Utah, Salt Lake City, UT, Phillip Blake, Department of Chemical Engineering, University of Arkansas, Fayetteville, AR, Braden Harbin, Microelectronics-Photonics, University of Arkansas, Fayetteville, AR, Aaron G. Russell, Department of Chemical Engineering, University of Akansas, Fayetteville, AR, Gyoung-Gug Jang, Ralph E. Martin Department of Chemical Engineering, University of Arkansas, Fayetteville, AR and Stuart Brune, Cell and Molecular Biology, University of Arkansas, Fayetteville, AR

Plasmon-enhancement of photovoltaics (PV) increases optical transition rate due to near-field plasmon absorption, and photocurrent generation from light trapping due to scattering. But enhancement is limited to plasmon resonant wavelengths. And increased thermalization and recombination, costly top-down fabrication, complex modeling, and distant semiconductor bandgaps prevent plasmons from sufficiently reducing PV cost per watt. We present a novel, bottom-up, lithographic method to fabricate ordered nanoarchitectures (see [figure 1]) to specifications set by a new exact solution to Maxwell's equations for electromagnetic interaction in these systems. These architectures scatter light at grazing angles, which maximizes light trapping. Lower particle densities reduce thermalization and recombination. In ordered nanoarchitectures, plasmons can couple with synchronous far-field radiation (SRPE) at infrared wavelengths near semiconductor bandgaps, producing >10-fold increases in local absorption and scattering that appear experimentally. Figure 2 [figure 2] shows spectra for single or random NP (green) vs. NP square lattice (blue) for Au and Si (inset). Au lattice SRPE peak is >102-fold higher than Si lattice SR peak, which lacks LSP, and >10-fold higher than Au NP LSP peak. The model identifies specifications to maximize semiconductor enhancement.

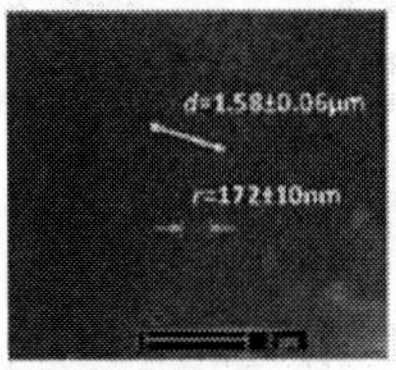

Fig. 1. Square Au NP lattices fabricated by novel, bottom-up method to specifications set by exact solution to Maxwell's equations.

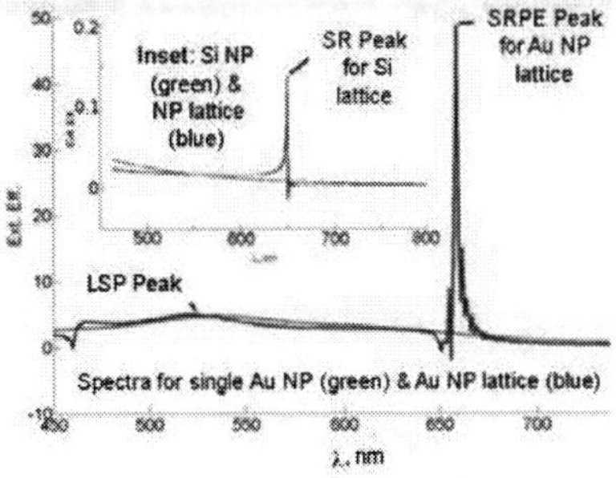

Fig. 2. Spectra for *single* or *random* Au NP (green) vs. Au NP *square lattice* (blue). **Inset**: Spectra for Si NP (green) vs. Si NP square lattice (blue). Au lattice SRPE peak at 650 nm is >10^2-fold higher than Si lattice SR peak, which lacks LSP, and >10-fold higher than Au NP LSP peak at 520 nm (no coherent scattering, i.e. SR). All NP have *r=70 nm*. Both lattices have *d*=650 nm.

See more of this Session: Nanomaterials for Photovoltaics III
See more of this Group/Topical: Topical 5: Nanomaterials for Energy Applications

(546c) Hot Electron Transfer From Semiconductor Nanocrystals

William A. Tisdale[1], Kenrick J. Williams[2], Brooke A. Timp[3], David J. Norris[1], Eray S. Aydil[1] and X.-Y. Zhu[2], (1)Chemical Engineering and Materials Science, University of Minnesota, Minneapolis, MN, (2)Chemistry and Biochemistry, University of Texas at Austin, Austin, TX, (3)Chemistry, University of Minnesota, Minneapolis, MN

In conventional semiconductor solar cells, absorption of photons with energies greater than the semiconductor band gap generate "hot" charge carriers that quickly "cool" before all of their energy can be captured – a process that limits device efficiency. Semiconductor nanocrystals (or quantum dots) have been touted as promising materials for photovoltaics because discretization of their electronic energy levels can slow down this cooling process, which might enable the extraction of photogenerated charge carriers before their excess energy is converted to heat.

In this talk, I will demonstrate sub-50 fs electron transfer from hot energy levels of PbSe nanocrystals to delocalized conduction band sates of TiO_2. In order to make these measurements, we developed the use of optical second harmonic generation for femtosecond time-resolved studies of interfacial charge separation. I will discuss the information we obtain from this technique as well as the effect of temperature, nanocrystal size, and surface chemistry. Additionally, I will show how ultrafast electron transfer excites coherent vibration of the first layer of TiO_2 surface atoms, whose collective atomic motions can be followed in real time.

See more of this Session: Nanomaterials for Photovoltaics III
See more of this Group/Topical: Topical 5: Nanomaterials for Energy Applications

(546d) Understanding and Controlling Organic-Inorganic Interfaces in Mesostructured Titania/Conjugated Polymer Materials for Photovoltaic Applications

Justin P. Jahnke[1], Shany Neyshtadt[2], Adu Rawal[1], Dan Huppert[3], Gitti L. Frey[2] and Bradley F. Chmelka[1], (1)Dept. of Chemical Engineering, UC Santa Barbara, Santa Barbara, CA, (2)Dept. of Materials Engineering, Technion, Haifa, Israel, (3)Dept. of Chemistry, Tel Aviv University, Tel Aviv, Israel

The use of surfactant-directed mesostructured inorganic materials allows for the preparation of oxide frameworks as continuous transparent films with high inorganic-organic interfacial contact and the inclusion of diverse functional guest species. For example, mesostructured (10 nm) titania films containing light-absorbing conjugated polymer guest species exhibit photovoltaic properties and may be synthesized by exploiting the compatible solution processabilities of the different molecular components that co-assemble into functional hybrid materials. Subsequent integration of these materials into photovoltaic devices leads to performances that depend on many factors, including the extent of interfacial contact between the conjugated polymer species and titania network. Such contact depends upon the molecular interactions among the titania framework, conjugated polymer species, and different structure-directing surfactants during synthesis and processing of the materials. Detailed molecular-level understanding of the interactions, mobilities, and proximities among the different functional components in the hybrid materials can be obtained through solid-state two-dimensional NMR and other techniques and correlated with macroscopic photo-current properties. In particular, judicious selections of the structure-directing surfactant and solvent species are demonstrated to promote increased contact between the conjugated polymer species and mesostructured titania network leading to improved photovoltaic device performance. Resulting insights allow mesostructured titania-conjugated polymer hybrid materials and device properties to be controllably modified and optimized.

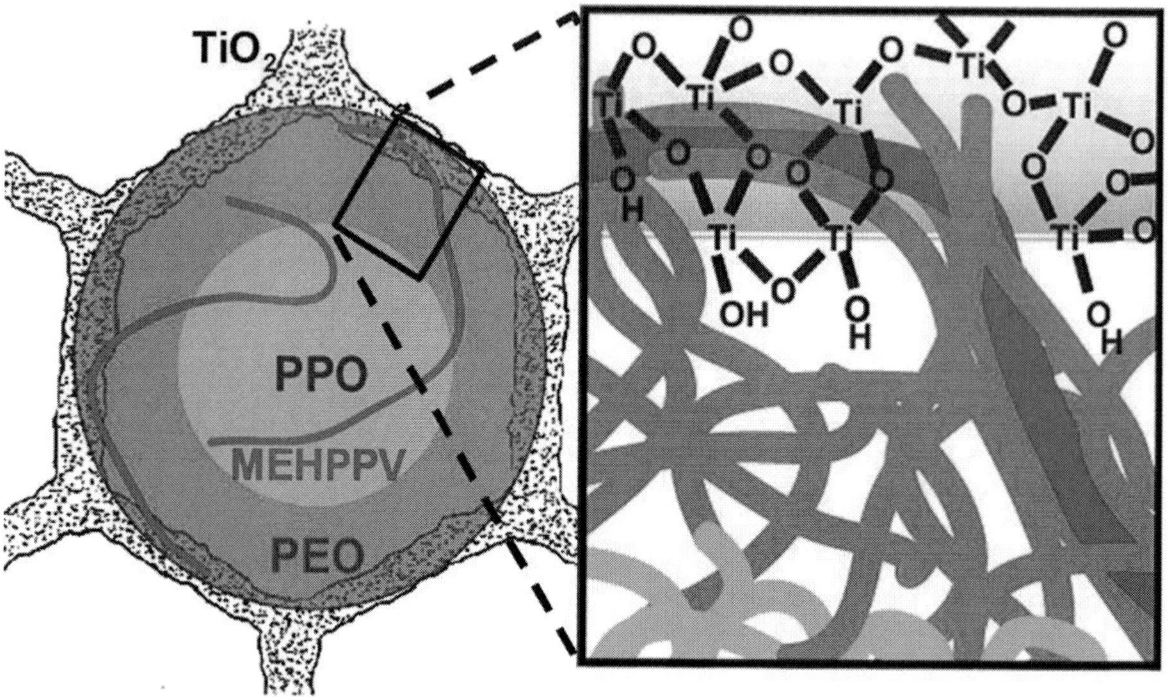

See more of this Session: Nanomaterials for Photovoltaics III
See more of this Group/Topical: Topical 5: Nanomaterials for Energy Applications

(546e) Electrochemical Studies On Porous Titania Electrodes for Photovoltaics

Sonia S. Mathew and Ilona Kretzschmar, Chemical Engineering, City College of New York, New York, NY

Dye-sensitized solar cells (DSSC) use porous titania photoanodes to convert solar energy to usable work. Traditionally, the titania electrode is a randomly connected network of titania nanoparticles. This configuration maximizes the surface area for dye-adhesion, which in turn maximizes the amount of energy harnessed. Our approach to improving the efficiency of the DSSC is to minimize recombination events, which is done by introducing order to the titania electrode. Our electrode is made of titania inverse opals with pore sizes on the order of 100s nm, compared to <20 nm in the DSSCs that are currently in the market. Our configuration avoids incomplete filling of the pores, which improves the electron transfer across the interface and reduces recombination events. We have used classic electrochemistry techniques like CV(cyclic voltammetry), EIS (electrochemical impedance spectroscopy), and CA (chronoamperometry) to characterize our electrodes. Previously, we have shown that charge density is inversely proportional with pore size in our system. This presentation discusses diffusion and electron transfer data for the inverse opal titania electrode.

See more of this Session: Nanomaterials for Photovoltaics III
See more of this Group/Topical: Topical 5: Nanomaterials for Energy Applications

(546f) Improving Photoactive Surface Area of Dye-Sensitized Solar Cells through Supercritical Fluid Dye Penetration

Fahd Rajab, David Loaring and Kirk Ziegler, Chemical Engineering, University of Florida, Gainesville, FL

While the benefits of solar cells are clear, these new technologies remain relatively expensive and inefficient. Dye-sensitized solar cells (DSSCs) are promising alternatives but photon collection and electron generation are dependent on the optical density of the film. Due to an inherent absence of an electric field within the nanoparticles, trapping affects charge carrier transport and back reactions such that diffusion coefficients and lifetimes react in opposition. This effect places an upper limit on the nanoparticle film thickness and the amount of absorbed light. Since film thickness cannot be increased in nanoparticle-based DSSCs, dye uptake becomes important in determining conversion efficiencies. Dye insertion through wet chemical or gas phase approaches are subject to capillary forces and condensation leading to inefficient coating of the titania nanostructures. Supercritical fluids (SCFs) cannot be condensed to a liquid phase minimizing these problems. Here we present the performance of DSSCs based on SCF impregnation of N-719 and N-749 (black dye). The better impregnation and coating of the titania may reduce back reactions, thereby enabling thicker titania films to be utilized. We also describe new methods to prepare thick, defect-free titania films and a non-destructive method to quantify the adsorption of dye.

See more of this Session: Nanomaterials for Photovoltaics III
See more of this Group/Topical: Topical 5: Nanomaterials for Energy Applications

(606a) Optimization of Power and Energy Densities in Electrochemial Double-Layer Capacitors

David B. Robinson, Energy Nanomaterials, Sandia National Laboratories, Livermore, CA

Supercapacitors use nanoporous electrodes to store large amounts of charge on their high surface areas, and use the ions in electrolytes to carry charge into the pores. Their high power density makes them a potentially useful complement to batteries. However, ion transport through long, narrow channels still limits power and efficiency in these devices. Proper design can mitigate this. Current collector geometry and electrolyte starvation must also be considered once this is done. Here, De Levie's model for porous electrodes[1] is applied to quantitatively predict device performance and to propose optimal device designs for given specifications. Effects unique to nanoscale pores are considered. Example design outlines for vehicle applications are proposed and compared. The optimal designs are currently not easily accessible given current technology for fabrication of electrodes and separators, but rapid progress in this area has been demonstrated recently [2-4]. Steering fabrication efforts toward these designs should allow for significant performance improvements, expanding opportunities for applications in electric vehicles, portable electronics, and power conditioning in electrical grids with distributed renewable sources.

[1] R. De Levie, Electrochimica Acta 8 (1963) 751-780.

[2] P. N. Bartlett, J. Marwan, Chemistry of Materials 15 (2003) 2962-2968.

[3] P. V. Braun, P. Osenar, V. Tohver, S. B. Kennedy, S. I. Stupp, Journal of the American Chemical Society 121 (1999) 7302-7309.

[4] S. Parida, D. Kramer, C. A. Volkert, H. Rosner, J. Erlebacher, J. Weissmuller, Physical Review Letters 97 (2006).

See more of this Session: Nanomaterials for Energy Storage I
See more of this Group/Topical: Topical 5: Nanomaterials for Energy Applications

(606b) Nanocapsulated Phase Change Materials Based On n-Octadecane Core and Conducting Polymer Shell

Sang Phil Park[1], Hyun Woog Ryu[1], Yeol Lee[1], In Woo Cheong[2], Jung-Hyun Kim[1] and Won-Gun Koh[1], (1)Chemical and Biomolecular Engineering, Yonsei Univ., Seoul, South Korea, (2)Applied Chemistry, Kyungpook Univ., Dae Gu, South Korea

Phase change materials (PCMs) play important role in regulating thermal energy by absorbing, storing, and releasing large amounts of latent heat over a defined temperature range during phase transition between liquid and solid. Usually, PCMs were encapsulated within polymeric shell to prevent leakage of liquid phase and various preparation methods of encapsulated PCM have been developed. However, most of developed techniques used micrometer-sized polymeric shells that do not have any function except containing PCMs. In this study, to overcome the limitations of current techniques, we prepared nanoencapsulated PCM by using n-octadecane as a model PCM and polypyrrole (Ppy) as a functional polymeric shell. PCM-PPy core-shell nanoparticles were obtained by miniemulsion polymerization using sodium lauryl sulfate (SLS) as a surfactant and dodecanol as a cosurfactant. Hydrogen peroxide (H2O2) and a trace of ferric chloride (FeCl3) were used as an initiator couple to carry out Fe3+-catalyzed oxidative polymerization, which occurred on the surface of nano-droplets. Average size of PCM-PPy core-shell nanoparticles was about 700 nm. The resultant core-shell structure was first confirmed by scanning electron microscope (SEM), where distinct core-shell morphology was observed after dissolving core PCM with petroleum ether dissolution. Heat capacity of resultant nanoencapsulated PCM was investigated with differential scanning calorimetry (DSC) and it was found that increasing the PCM content lead to enhanced heat capacity (13 J/g ~ 22 J/g). Electrical conductivity is analyzed by 4-probe conductivity meter and it was observed that the doped particles showed a high conductivity in the dry state. In summary, nanoencapsulated PCM was successfully prepared using miniemulsion process and using Ppy as a polymeric shell endowed PCM nanoparticles with conductive property as well as heat storage property, which will open more opportunity for various applications of PCMs.

See more of this Session: Nanomaterials for Energy Storage I
See more of this Group/Topical: Topical 5: Nanomaterials for Energy Applications

(606c) Nanoscale Organic Hybrid Electrolytes for Lithium Metal Batteries

Jennifer L. Nugent, Surya S. Moganty and Lynden A. Archer, Chemical Engineering, Cornell University, Ithaca, NY

Nanoscale hybrid electrolyes have been synthesized that show promise for use in lithium metal batteries. Nanoparticle organic hybrid materials (NOHMs) were created with silica nanocores covalently bonded to polyethylene glycol oligomers (the corona). NOHMs were purified by precipitation in ethanol to remove free organic material, then doped to 1 M LiTFSI in the organic phase. Unlike traditional composite polymer electrolyes, NOHMs based electrolytes are homogenous with evenly distributed, non-agglomerated nanoparticles [1]. Suspension of the NOHMs in free polyethylene glycol dimethyl ether (PEGDME) with 1 M LiTFSI results in a family of materials with properties ranging from liquid-like to solid-like. At a critical NOHMs fraction, the blend is jammed and the electrolytes display soft-glassy rheology. This jamming phenomenon results in a 6 order of magnitude increase in storage modulus with only a fractional decrease in ionic conductivity. The electrolyte properties could be further altered by changes to the dilutant, core size and chemistry, and corona length and chemistry. This level of mechanical tunability may be used to study the effect of modulus and yield stress on dendrite growth upon cycling of lithium metal cells. In addition, as dendritic lithium growth initiates from uneven mass transport pathways of Li+ to the electrode interface, the length scale of homogeneity in the transporting organic phase and its effect on lithium plating may be studied by varying spacing between the inorganic cores. [1] J. L. Nugent, S. S. Moganty, and L. A. Archer. Nanoscale Organic Hybrid Electrolytes. Adv. Mater. In Press.

See more of this Session: Nanomaterials for Energy Storage I
See more of this Group/Topical: Topical 5: Nanomaterials for Energy Applications

(606d) Template-Free Synthesis of Fe_3O_4 Nanoparticles and Their Performance as Anode Materials in Lithium-Ion Batteries

Zichao Yang, Jingguo Shen, Surya S. Moganty and Lynden A. Archer, Chemical and Biomolecular Engineering, Cornell University, Ithaca, NY

The growing demand for small, light-weight, and inexpensive power sources for portable electronics and automotive uses is driving interest in lithium-ion battery (LIB) technologies that exceed performance of today's batteries. Metal oxides are attractive as anode materials for LIBs due to their high lithium intercalation capacity and generally high natural abundances. Magnetite, Fe_3O_4, is a particularly promising candidate because of its high theoretical capacity of 924 mAh/g (cf. 372 mAh/g for currently used graphite), low cost, and low environmental footprint. A major challenge that limits development of superior anode materials based on this material is the pulverization problem, namely, the breakdown of electrical contact between particles of the active material produced by the large, cyclic volume changes which accompany repeated lithium insertion/extraction. Herein, we report results from a systematic study of hollow Fe_3O_4 nanoparticles synthesized via a template-free solvothermal method. Temporal XRD and TEM characterization indicate that the growth follows an inside-out Ostwald ripening mechanism, that spontaneously produces hollow nanostructures irrespective of the particle chemistry. Electrochemical cyclic voltammetry studies show that both hollow and filled Fe_3O_4 nanostructures reversibly intercalate lithium at ~0.9 and 1.2 V vs. Li/Li^+. When integrated in the LIB anode, we find a systematic improvement in performance as the internal void space is increased. We also study the effects of creating a carbon-Fe_3O_4 nanocomposite by compositing Fe_3O_4 with polyacrylonitrile (PAN) followed by calcination to convert PAN to carbon. This procedure was found to consistently improve the electrochemical performance of the particles. The origin of this improvement is investigated using post-mortem TEM and nanoindentation studies, as well as in-situ characterization of the lithium ion diffusion by impedance spectroscopy.

See more of this Session: Nanomaterials for Energy Storage I
See more of this Group/Topical: Topical 5: Nanomaterials for Energy Applications

(606e) Chemically Driven, Carbon Nanotube-Guided Thermopower Waves

Wonjoon Choi[1], Seunghyun Hong[2], Joel T. Abrahamson[3], Jae-Hee Han[3], Changsik Song[3], Nitish Nair[3], Seunghyun Baik[2] and Michael S. Strano[3], (1)Department of Mechanical Engineering, Massachusetts Institute of Technology, Cambridge, MA, (2)Advanced Institute of Nanotechnology, Department of Energy Science and School of Mechanical Engineering, Sungkyunkwan University, Suwon, South Korea, (3)Department of Chemical Engineering, Massachusetts Institute of Technology, Cambridge, MA

Coupling an exothermic chemical reaction with a nanotube or nanowire possessing a high axial thermal conductivity creates a self-propagating reactive wave along its length. Such waves are realized using a 7-nm cyclotrimethylene-trinitramine annular shell around a multi-walled carbon nanotube and are amplified by more than 10^4 times the bulk value, propagating more than 2 m/s, with an effective thermal conductivity of 1.28 ± 0.2 kW/m/K at 2860 K. This wave produces a concomitant electrical pulse of disproportionately high specific power, as large as 7 kW/kg, that we identify as a thermopower wave. Here, thermally excited carriers flow in the direction of the propagating reaction with specific power that scales inversely with size. The reaction also produces an anisotropic pressure wave of decomposition gases of exceedingly high total impulse per mass of 300 N-s/kg and specific impulse per total mass (5.5 s/μg) suggesting possible applications to micro-propulsion and actuators.

See more of this Session: Nanomaterials for Energy Storage I
See more of this Group/Topical: Topical 5: Nanomaterials for Energy Applications

(606f) Development of Nanoporous Si/Graphene Nanomaterial for High Performance Lithium Batteries

Rhet Joseph De Guzman[1], Jinho Yang[2], Mark Cheng[2], Steven Salley[1] and K. Y. Simon Ng[1], (1) Department of Chemical Engineering and Materials Science, Wayne State University, Detroit, MI, (2) Department of Electrical and Computer Engineering, Wayne State University, Detroit, MI

The interest in developing higher energy capacity and longer cycle life rechargeable lithium batteries is steadily increasing with the realization of its potential capabilities especially in electric vehicles. Recently, silicon-based anode materials are being explored as an anode material due to its low discharge potential and its theoretical capacity (highest known, 4,200 mAh/g) which is more than 10 folds of the traditional graphite anodes. However, due to the increased accommodated lithium ions, silicon's volume varies by 400% causing pulverization and loss of electrical contact. This study explores the use of electrochemically etched nanoporous silicon to maximize surface area for lithium alloying and to allow spaces for volume expansion. The material is then coated with graphene overlayers to enhance capacity retention and cyclability thru stabilization of the solid-electrolyte interface.

See more of this Session: Nanomaterials for Energy Storage I
See more of this Group/Topical: Topical 5: Nanomaterials for Energy Applications

(655a) Nitrides for Supercapacitor Applications

Prashanth Jampani Hanumantha and Prashant Kumta, Chemical Engineering, University of Pittsburgh, Pittsburgh, PA

The need for a more sustainable solution to the energy needs of the world has been felt and is driving various new technologies including lithium ion batteries, supercapacitors and fuel cells. Supercapacitors hold a unique place among these energy storage devices on account of their high power densities and long cycleability.

Nanocrystalline Vanadium nitride (VN) has been shown to be a promising supercapacitor material with capacitances as high as 1300 F/g. The capacitance however drops in a few cycles. In this work, we show that doping VN with various other transition metals leads to considerable increase in stability. This improvement in performance has been shown at various scan rates and various loadings.

See more of this Session: Nanomaterials for Energy Storage II
See more of this Group/Topical: Topical 5: Nanomaterials for Energy Applications

(655b) Carbon Coated Co3O4 as High Power Anode for Lithium Battery Applications

Jayaprakash Navaneedhakrishnan, Surya Sekhar Moganty and Lynden A Archer, School of Chemical and Bio-Molecular Engineering, Cornell University, Ithaca, NY

Today's world of modernization and miniaturization lays greater emphasis on more power from smaller and lighter battery packs. Among the competing systems, rechargeable lithium batteries show the highest energy density, due to the high reducing power of lithium that leads to large cell voltage. The limitations in the overall performance of Li-ion batteries depend on the intrinsic performances of the materials (anode, cathode and electrolyte) and the technology aspects (material processing, electrode fabrication and battery conception), taking into consideration of the environment of each material in the complete cell. Tremendous amount of work has been done worldwide in these two directions over the last twenty years. The active anode material of a secondary Li-ion battery is a host material into / from which lithium can be reversibly inserted / extracted over a composition range. Currently, carbonaceous materials are widely used as anodes in Li-ion batteries because of their low cost, even discharge characteristics and low operating voltage. Generally, the commercialized carbonaceous anodes though have found widespread applications, problems like the high irreversible capacity, swelling behavior or failure of the electrode through intercalation of propylene carbonate (PC) based electrolyte and the formation of surface electrolyte interface (SEI) layer, remain as severe unsolved problems till date. Disruption of this SEI layer in conventional Li-ion batteries is a primary source of capacity and power fade. Recent advances in the synthesis of tin (Sn), silicon (Si), nickel (Ni), cobalt (Co), and their corresponding oxides as replacements for carbon-based anodes have resulted in batteries with higher specific capacity and enhanced cycle life. Based on this ground, we have made an attempt to exploit carbon coated Co3O4 as anode for high power lithium battery applications. Herein, hydrothermal method has been used to synthesize monodisperesed spherical Co3O4 particles and subsequent carbon coating using glucose as the carbon source. Experimental: Monodispersed, spherical Co3O4 particles was synthesized by the ammonia assisted hydrothermal method at 180 °C and the carbon coating to the synthesized Co3O4 particles was done using glucose as the carbon source at the same temperature. Electrochemical properties of the synthesized C- Co3O4 material was evaluated using a 2032 type coin cell. The cell was assembled in an argon-filled glove box and charged and discharged between 0.5 and 3.0 V at room temperature. Results and discussion Figure 1 shows the surface morphology and the cyclic voltammetry studies of the C-Co3O4 material. It is interesting to note that the ammonia assisted hydrothermal synthesis methodology has produced spherical Co3O4 particles with well defined grain boundary. Carbon coating of ~10nm is obvious from the TEM images obtained. When deployed as anode in lithium-ion batteries, the C-Co3O4 has delivered an excellent cycle life as well as rate capability behavior. The various physical and electrochemical parameters of the synthesized compounds are discussed in detail.

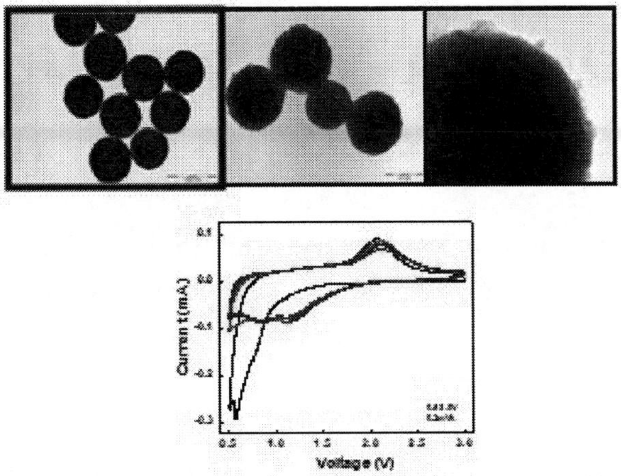

Figure 1. Surface morphology (TEM) and electrochemical behavior (CV) of the pristine and C-Co_3O_4 materials.

See more of this Session: Nanomaterials for Energy Storage II
See more of this Group/Topical: Topical 5: Nanomaterials for Energy Applications

(702a) Novel Quaternary Pt-Ru-Ni-Ti Alloy System for Direct Methanol Fuel Cell Electrocatalyst

Karan Kadakia[1], Moni Kanchan Datta[2] and Prashant Kumta[1], (1)Chemical Engineering, University of Pittsburgh, Pittsburgh, PA, (2)Mechanical Engineering & Materials Science, University of Pittsburgh, Pittsburgh, PA

Direct methanol fuel cells (DMFC) are of interest as suitable power sources for consumer electronic devices as well as remote and auxiliary power units for transportation. The main advantage of DMFC is the use of liquid methanol rather than gaseous hydrogen as the operating fuel used in the hydrogen fueled proton exchange membrane fuel cell (PEM). The relatively sluggish kinetics of the electrochemical reactions occurring on the catalyst layers at the electrode surface combined with the methanol crossover to the cathode compartment reduces the power density of the DMFC. There is therefore a critical need to research and develop new and improved electro-catalysts with higher catalytic activity in order to enhance the efficiency and power density of DMFC, while lowering the precious metal loadings to realize the much desired cost savings to match the performance. Currently, Pt-Ru alloy of nominal composition Pt-50at.% Ru is considered the most suitable catalyst composition yielding the best catalytic activity and is hence widely recognized as the anode catalyst for use in DMFCs. Improved anode catalysts can be obtained by synthesizing high surface area Pt-Ru based alloy catalysts combined with engineered proper alloy design by modifying the chemical composition of the alloy, and doping or alloying with other elements onto the catalyst structure. It has been proposed that addition of other alloying elements could reduce material loss during cycling and particle ripening. Ni-Zr and Ni-Ti have been previously shown to be electrochemically active and stable under anode conditions. In the present work, we report bulk synthesis of nanocrystalline Pt-Ru-Ni-Ti based quaternary alloys, for use as DMFC anodes, by a facile wet chemical synthesis route. X-ray results show complete alloying and formation of phase pure Pt(Ru,Ni,Ti) solid solution, and also nanocrystalline nature of the alloy. Alloy materials have been shown to have high BET surface areas (> 200 m^2/g). Cyclic voltammetry and polarization results indicate that the nano-crystalline Pt-Ru-Ni-Ti alloy of suitable compositions, possessing good chemical homogeneity, reveals excellent catalytic activity, demonstrating the potential of the Pt-Ru-Ni-Ti based electrocatalysts for direct methanol fuel cells.

See more of this Session: Nanomaterials for Hydrogen Production and Fuel Cells I
See more of this Group/Topical: Topical 5: Nanomaterials for Energy Applications

(702b) Stability and ORR Activity of Nitrogen Functionalized Ordered Mesoporous Carbon Supports

Sujan Shrestha and William E. Mustain, Department of Chemical, Materials, and Biomolecular Engineering, University of Connecticut, Storrs, CT

The most popular catalyst for low temperature proton exchange membrane fuel cells (PEMFC) is platinum (Pt) nanoparticles supported on carbon [1]. One of the major issues concerning Pt nanoparticles is the very slow kinetics of oxygen reduction reaction (ORR) which limits overall performance of PEMFC [2]. Moreover, catalytic activity and thereby, fuel cell performance is lost overtime due to loss of electrochemically active surface area (EASA) at the cathode caused by particle agglomeration, dissolution and redeposition, and active site contamination [3, 4].

One of the ways to improve stability and performance of the Pt electrocatalyst is to introduce surface functional groups into the carbon support [3]. We introduce nitrogen functionalized ordered mesoporous carbon as support for Pt electrocatalyst. The nitrogen functionalized ordered mesoporous carbon (NOMC) support has narrow pore size distribution and allows easy tunability of its pore size [5]. Here we characterize the cycleability, Pt particle distribution, and ORR activity of Pt deposited NOMC, which shows promise as next generation ORR catalyst.

References

1. L. Zhang, J. Zhang, D. P. Wilkinson, and H. Wang. Journal of Power Sources, 156 (2006) 171-182. 2. K. C. Neyerlin, W. Gu, J. Jorne, and H. A. Gasteiger. J. Electrochem. Soc., 153 (2006) A1955-A1963. 3. S. Zhang, X. –Z. Yuan, J. N. C. Hin, H. Wang, K. A. Friedrich, and M. Schulze. Journal of Power Sources, 194 (2009) 588-600. 4. Y. Shao-Horn, W. C. Sheng, S. Chen, P. J. Ferreira, E. F. Hobly, and D. Morgan. Topics in Catalysis, 46 (2007) 285-305. 5. D. Zhao, J. Feng, Q. Huo, N. Melosh, G. H. Fredrickson, B. F. Chmelka, and G. D. Stucky. Science, 279 (1998) 548-552.

See more of this Session: Nanomaterials for Hydrogen Production and Fuel Cells I
See more of this Group/Topical: Topical 5: Nanomaterials for Energy Applications

(702c) A Molecular Dynamics Simulation Study of Hydrated Sulfonated Poly (Ether Ether Ketone) for Application to Polymer Electrolyte Membrane Fuel Cells

Seung Soon Jang, Materials Science and Engineering, Georgia Institute of Technology, Atlanta, GA and Giuseppe F. Brunello, School of Materials Science and Engineering, Georgia Institute of Technology, Atlanta, GA

S-PEEK has been known as one of the promising candidates for fuel cell application since it has high thermo-oxidative stability and extraordinary mechanical properties as well as high cost effectiveness. However, the nanoscale structure and transport properties for S-PEEK have not been thoroughly reported at molecular level. Through this study, we have investigated the the nanophase-segregated structure of hydrated S-PEEK membrane using atomistic molecular dynamics simulation method. We will present the effect of water content and temperature on the nanophase-segregation and water transport of the hydrated S-PEEK membrane. The simulation results for the S-PEEK membrane are discussed in comparison with the experimental data available in the literatures as well as those for the Nafion membrane.

See more of this Session: Nanomaterials for Hydrogen Production and Fuel Cells I
See more of this Group/Topical: Topical 5: Nanomaterials for Energy Applications

(702d) Molecular Dynamic Simulations of the Effect On the Hydration of Nafion at the Presence of Nanoparticle

Parag Adhangale, Shenghong Zhang and David Keffer, Department of Chemical and Bimolecular Engineering, University of Tennessee, Knoxville, TN

Nafion possesses good chemical, thermal and mechanical properties as a proton conductive membrane in proton exchange membrane (PEM) fuel cells with relatively high proton conductivity. As such, Nafion is the current industry standard PEM. Nafion performs best under the high humidity conditions at moderate temperature. From the perspective of the catalyst in a fuel cell, one would like to increase the operating temperature of the fuel cell (1) to increase the activity of the catalyst, (2) decrease the susceptibility to catalyst poisoning by CO and (3) reduce the amount of catalyst required. However the performance of Nafion falls rapidly as the temperature rises over 80 ^{0}C due to dehydration of the membrane, which results in a drastic drop in conductivity.

Consequently, there has been a tremendous effort to develop membranes that retain moisture at higher temperatures (90-120 ^{0}C), with the understanding that better retention of water will result in higher conductivities in the PEM at elevated temperatures. The investigation of new high temperature membranes include composite membranes in which Nafion contains an hydrophilic inorganic material either in the form of an interpenetrating nanostructured matrix or as dispersed nanoparticles.

Various materials such as SiO_2, ZrO_2, TiO_2 and Pt have been evaluated over the past two decades for their ability to improve the performance of fuel cells at elevated temperatures. The results showed that the weight fraction of nanoparticle as well as the morphology, surface orientation and modification of the nanoparticles have important effects on the ability to retain moisture and on the resulting conductivities. These improved properties were reasoned to the strong interfacial interactions between the polymer matrix and the filler surface, as opposed to the conventional composites. However, a clear molecular-level understanding of such effects does not currently exist.

In this study, molecular dynamic simulations were applied to investigate the role of the various nanoparticles dispersed in Nafion polymers in the retention of moisture and promotion of high conductivities. Simulations in the isobaric-isothermal ensemble are used to predict composite membrane densities and water content as a function of temperature and relative humidity.

See more of this Session: Nanomaterials for Hydrogen Production and Fuel Cells I
See more of this Group/Topical: Topical 5: Nanomaterials for Energy Applications

(702e) Aligning the Ionic Nano-Channels in Nafion Membrane

Yuxin Wang[1], Shixiong Zhao[1] and Li Xu[2], (1)School of chemical engineering, United state key lab of chemical engineering, Tianjin University, Tianjin, China, (2)School of Chemical Engineering, Tianjin University, Tianjin, China

Nafion membrane is the most used proton exchange membrane in PEM fuel cells and PEM water electrolysis, because of its high ionic conductivity and excellent chemical, thermal and mechanical properties. It is known that Nafion, as well as other similar proton exchange membranes, has conductive channels consisted of sulfonate ionic nano-clusters randomly distributed in nonconductive polymer matrix. If the ionic conductive channels in the membrane can be aligned, for instance by the mechanism of polar-polar interaction under electric field, the ionic conductivity of the membrane should increase, which could lead to significant performance improvement in fuel cell and water electrolysis. However, the nano-structure alignment in the Nafion membrane proved to be not straightforward. Although the alignment of ionomer phase in blend polymer membranes was achieved by exerting electric field, the addition of a nonconductive polymer led to decreased conductivity. Electric field was shown to be not effective if casting Nafion membrane with usual Nafion solutions. Reported improvement by applying electric field to solid Nafion membrane was not repeatable. In this presentation, we report our recent finding that electric field become effective in aligning the conductive channels in Nafion membrane when using a specific binary organic solvent to make Nafion casting solution. The trans-plane proton conductivity of the electrocasting Nafion membrane increases by 23% as compared to that of non-electrocasting membrane. Interestingly, the binary solvent casting membrane showed conductivity increase by 11% even without exerting electric field. Properties of the binary solvent membrane casting solution and other features of the electrocasting Nafion membrane will also be presented. Factors affecting the alignment of ionic channels in Nafion membrane will be discussed.

See more of this Session: Nanomaterials for Hydrogen Production and Fuel Cells I
See more of this Group/Topical: Topical 5: Nanomaterials for Energy Applications

(702f) Zeolite – Protic Ionic Liquid Composites: Preparation, Characterization and Evaluation of Ion Conduction Properties

Vladimiros Nikolakis, Spyros Ntais, Anastasia Maria Moschovi, Stylianos Neophytides, Vassilis Burganos and Vassilis Dracopoulos, FORTH/ ICEHT, Patras, Greece

Fuel Cells (FCs) have been recognized as one important device for efficient transformation of chemical energy to electricity. Among the different FC types those of Polymer Electrolyte Membranes (PEMs) have received significant attention. In such FCs, a membrane that ideally enables only proton transport is the barrier between the anode and cathode. Nafion, which is usually used for this purpose, requires relatively low operating temperatures (<100°C) in order to maintain the necessary levels of hydration for proton conduction. However, the low operating temperature has two significant drawbacks: 1. It is not easy to recover the waste heat; and 2. The Pt catalyst of the electrodes is very sensitive to CO poisoning. To address these issues many researchers are trying to develop High Temperature Polymer Electrolyte Membranes (HTPEMs), with operating temperatures of up to 200°C. Unfortunately, the performance of HTPEMs is often limited by the membrane fuel crossover and the loss of hydration which results in the loss of proton conductivity. The incorporation of hydrophilic zeolite nanocrystals in PEs has been suggested as a possible alternative for increasing membrane water content at elevated temperatures and decreasing fuel cross-over.

Zeolites conduction is primarily due to cation migration[1], and is significantly enhanced in the presence of water. However, even in that case the maximum conductivity values reported are of the order of ~10^{-4} $S.cm^{-1}$ [2]. These values are relatively low indicating that the presence of zeolites is expected to have a negative effect on the overall PEM proton conduction. To improve zeolite conductivity several groups have synthesised and tested acid functionalized mesoporous materials and zeolites[3].

Ionic liquids (ILs) are salts with low melting points that are characterized by their enhanced thermal stability, non-volatility and non-flammability. It has recently been shown that protic ILs can undergo hydrogen oxidation and oxygen reduction reactions in the absence of water molecules, thus they have the potential to be used as proton carriers in Fuel Cells operating under anhydrous conditions[4].

The goals of this work are, initially, to study the encapsulation of a protic IL in zeolite NaY, and to understand its effect on zeolite conduction properties. Such composite materials might become one constituent of a composite membrane for HTPEM fuel cells. H-3-methylimidazolium bis (trifluoromethanesulfonyl) imide (HMITFSI) was selected as the protic IL of this study. Composites with HMITFSI/ Zeolite NaY weight ratios (IL/Z) between 0.03 and 1 were prepared. N_2 physisorption measurements have shown that the micropore surface area gradually decreases with IL loading, indicating that HMITFSI enters the zeolite pores. Furthermore, the water uptake of the zeolite crystals decreased from ~9% at 150°C for zeolite NaY to ~2.5% for the sampe with IL/Z ~1. Analysis of X-ray diffraction data and FT-Raman spectra show the formation of NaTFSI which indicates that the entrapment of HMITFSI is also accompanied by an ion-exchange reaction. The ionic conductivity of the composites was measured using impedance spectroscopy. At dry conditions it increased with temperature and with IL/Z ratio. The maximum conductivity was ~2.5 $mS.cm^{-1}$ at 150°C for the sample IL/Z=1. This value is about 4 orders of magnitude higher than that of NaY powder and about 1 order of magnitude lower than that of the pure IL. The presence of ~4.2 kPa of water primarily enhanced the conductivity of the samples with low IL/Z values. The Effective Medium Theory was also used to estimate the zeolite/ionic liquid conductivity at the grain scale as inferred from conductivity measurements at the macroscopic scale. The results will be analysed with respect to the possible proton transfer mechanisms and with respect to the potential use of such composites in Fuel Cell devices.

---

[1] D. C. Freeman, D. N. Stamires, J. Chem. Phys., 35 (3), (1961), 799-806.

[2] D. N. Stamires, J. Chem. Phys. 36 (1962), 3174.

[3] (a) J. C. McKeen, Y. S. Yan, M. E. Davis, Chem. Mater., 20, (2008), 3791-3793, (b), J. C. McKeen, Y. S. Yan, M. E. Davis, Chem. Mater. 20, (2008), 5122-5124, (c) R. Marschall, I. Bannat, J. Caro, M. Wark, Microp. Mesop. Mater., 99, (2007), 190-196, (d) Yu-Feng Lin, Chuan-Yu Yen, Chen-Chi M. Ma, Shu-Hang Liao, Chia-Hsun Lee, Yi-Hsiu Hsiao, Hong-Ping Lin, J. Power Sources, 171, (2007) 388–395, (e) J. D. Halla, M. Mamak, D.E. Williams, G.A. Ozin, Adv. Funct. Mater., 13(2), (2003), 133-138.

[4] (a) **Md. A. B. H. Susan, A. Noda, S. Mitsushima, M. Watanabe, Chem. Commun., (2003), 938-939, (b) H. Nakamoto, M. Watanabe, Chem. Commun., (2007), 2539-2541.**

See more of this Session: Nanomaterials for Hydrogen Production and Fuel Cells I
See more of this Group/Topical: Topical 5: Nanomaterials for Energy Applications

PVA-Calcium Oxide Membrane for Direct Methanol Fuel Cell

Mat,N.C, Yusof,M., Chiong,K.K.C

Department of Chemical Engineering and Energy Sustainability, Faculty of Engineering,
University Malaysia Sarawak, 94300 Kota Samarahan, Sarawak, Malaysia.

ABSTRACT

The large scale utilization of Nafion® in Direct Methanol Fuel Cell(DMFC) is still constrained by high methanol permeability and high temperature dehydration. The methanol crossover in DMFC can be minimized by using pervaporation membrane such as Poly-vinyl alcohol (PVA) due to higher water selectivity over alcohol. Incoporation of hygroscopic oxide particle into polyelectroylte membrane for DMFC has draw interest among researchers as it has been reported being able to improve the water retentation at the elevated temperature attributed to the excellences thermal stability and high swelling resistances. In this work, new composite membrane comprised of poly-vinyl alcohol (PVA) and Calcium Oxide were fabricated by using sol-gel method. The objectives of this study were to investigate the effect of CaO addition to the PVA conductivity, water uptake, and methanol permeabilty properties. From the result of the study, CaO have the profound effects on the PVA conductivity, water uptakes and methanol permeability properties thus show a potential to be used for direct methanol fuel cell applications.

1. INTRODUCTION

The direct methanol fuel cell (DMFC) has been highlighted as a possible energy converter for portable electronic devices, transportation and stationary power application. Nafion® (Du Pont) is widely used as proton conducting electrolyte membrane due to electrochemical and mechanical stabilities as well as high conductivity. However, the large scale utilization of Nafion® is DMFC is still constrained by high cost and high temperature dehydration. Nafion® also is very permeable to methanol and that leads to methanol crossovers to the oxygen/air cathode which eventually reduce the overall fuel cell efficiency of the system. Therefore, the development of low methanol permeability and suitable proton conductivity at high temperature is crucial [1, 2]. The methanol crossover rate in DMFC can be minimized by using pervaporation membrane such as Poly-Vinyl Alcohol (PVA). PVA has been widely used in pervaporation- based alcohol due to the higher water selectivity over alcohol [3]. However, the application of PVA in DMFC is still hindered by poor proton conduction and mechanical properties [4]. Incorporation of inorganic fillers such as Montmorillonite (MMT) into PVA has improved the thermo mechanical and ionic conductivity properties. In addition, methanol crossover rates effectively reduced owing to the layered structure and high aspect ratio of the inorganic fillers [5]. Inclusion of hygroscopic oxide particle in polyelectrolyte membrane for DMFC application has draw interest among researchers as it has been reported being able to improve the water retention at the elevated temperature attributed to the excellence thermal stability and high swelling resistances. Moreover, strong interface adhesion between inorganic hygroscopic particles and organic polymeric matrix could suppress the methanol diffusion from anode to cathode thus the overall efficiency could be improved [6]. Calcium Oxide (CaO) is a hygroscopic and inorganic proton conducting material [7]. CaO has been widely used in construction industries and catalyst in biodiesel transesterification [8]. CaO also have been used as solid fuel components for DMFC due to CO_2 absorption properties at the anode thus improvement on long term DMFC operation could be achieved

2. EXPERIMENTAL

2.1 Materials

Poly(vinyl)alcohol(PVA),Tetraethylorthosilicate(TEOS) Phosphotungstic acid (PWA) and ethanol (all chemicals are purchased from Sigma Aldrich) were used to prepare polyelectrolyte membrane. All chemicals were used without further purification.

2.2 Membrane preparation

PVA were dissolved in deionized water with continous stirring at 70°C. The various weight ratio of CaO (0 wt%, 2.5wt%, 5wt% and 7.5wt%) was then added to this PVA solution and the mixture was stirred continuously until a homogenous solution was obtained. PVA and CaO solutions were prepared separately before added together. The PVA solutions were then prepared with subsequent treatment in phosphotungstic acid (PWA) solution and the mixture was stirred at 70°C for 4 hours. On the other hand, CaO solution was prepared by adding the TEOS mixture through *sol-gel* reaction. The TEOS mixture was prepared by mixing ethanol and TEOS in the molar ratio of 4:1 and then solution was stirred at 90°C for 4 hours. This solution was then added to the PVA solution and the membranes were cast by pouring the solution onto glass plate. The cast solutions were allowed to dry in air at room temperature. After that, the fully dried membranes were peeled away from the glass plate and heated in an oven at 80°C for 2 hours.

2.3 Fabrication of membrane electrode assembly (MEA) and evaluation of single cell performance

The electro catalysts were prepared and deposited using a direct dispersion method. The cathode catalyzed layer consisted of 4.0 mg cm $^{-2}$ Pt black dispersed on carbon black and bound together with 30 wt. % PTFE were spread on this diffusion backing layer. The anode was constructed by using 4.0 mg cm $^{-2}$ Pt-Ru The geometrical area of the electrodes was 6.0cm^2. The fabricated membrane was sandwiched between two electrodes by hot pressing method. The single cell performance experiments were performed at room temperature and at atmospheric air pressure.

2.4 Water uptake

Water uptake of membranes were performed by drying the samples of known weight at 100°C for 24 hours, immersing them in deionized water and equilibrating it for 2 days. The weight of swollen membranes W_{wet}) was measured carefully after carefully removing the liquid from both surfaces with filter paper. Weight of dry membrane (W_{dry}) was determined after being completely dried after 100°C for 12 hours. The water uptake tests calculated by following equations:

$$\text{Water uptake (\%)} = = \frac{W_{wet} - W_{dry}}{W_{dry}} x100$$

2.5 Swelling test

The swelling ratio was determined by measuring the change of the membrane area after equilibrating the membrane in methanol for 24 hours. The swelling ratio was determined by using this following equation

$$\text{Swelling (\%)} == \frac{A_{wet} - A_{dry}}{A_{dry}} x100$$

where A_{dry} and A_{wet} are the area of dry and wet samples.

2.6 Methanol permeability test

The methanol permeability test of Nafion 117, PVA and fabricated PVA-CaO composite membrane was determined by using mini fuel cell test systems which connected with two reservoirs A and B. All membranes were first pretreated with 2% hydrogen peroxide solutions and 0.5 M sulfuric acid solutions. Membranes were kept in distilled water for overnight. Membrane areas of 36 cm^2 were sandwiched between cathode and anode catalyzed layer. 200 ml of 1.0 M methanol solution were added to the reservoir and same amount of distilled water added in reservoir B in order to maintain equal hydrostatic pressure and concentration on both side of membrane. The potential differences were measured by using multimeter. The concentration of methanol at reservoir B at time t, C_B was estimated by using Eq (1)

$$E = E^0 - \left(\frac{RT}{F}\right) \ln \left[\frac{C_B(t)}{\theta_{ad} C_{H^+}}\right] \quad \text{Eq (1)}$$

where E0 is the open potential circuit, F is the Faraday constant and θ_{ad} is the surface coverage by the absorbed species [10]. The experiment was conducted at 70°Cwhile permeability of methanol was determined by using Eq (2) [6] , where C_A is the methanol concentration in reservoir A, V_B is the volume of the reservoir B, A and L is the membrane area and thickness, D and K are the methanol diffusivity and partition coefficient The D value determined from the . The membrane permeability P is also determined from the product DK

$$C_B(t) = \frac{A}{V}\frac{DK}{L}C_A(t - t_o) \quad \text{Eq(2)}$$

2.6 Conductivity test

The proton conductivity test was carried out by using electrochemical impedance spectroscopy technique (EIS). The detailed of this technique has been described elsewhere [10]. Membranes were cut into 2.0 cm x 2.0cm before soaked in water for 2 hours. The conductivity,(σ) of the membrane samples was determined from σ=L/RA where L is the thickness of the membrane samples , A is the area of the membrane and R is the membrane resistance which derived from the impedance data.

3.0 RESULTS AND DISCUSSIONS

3.1 Water uptake test

Table 1 shows the water uptake percentages of Nafion 117, PVA and fabricated PVA-CaO composite membrane at room temperature. Lowest water uptake was observed in Nafion 117. The water uptake percentages increased with CaO.

Membrane	Water uptake (%)
Nafion 117	28
PVA	85
PVA +2.5%CaO	110
PVA+5.0%CaO	137
PVA+7.5%CaO	154

Table 1: Water uptake percentages for Nafion 117, PVA and the fabricated membrane.

3.2 Swelling properties

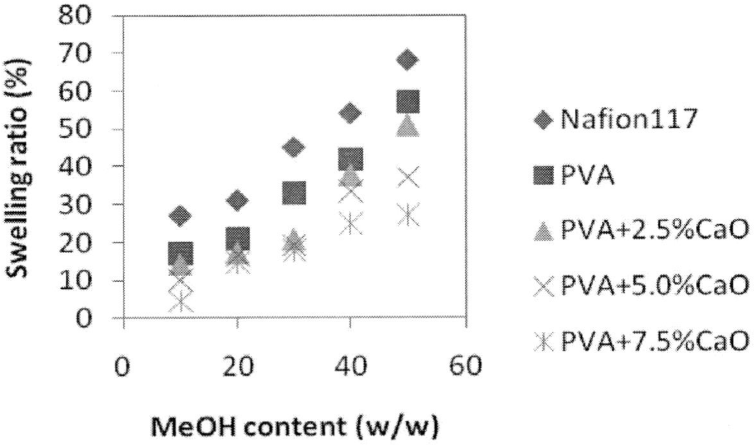

Figure 2: Methanol swelling ratio of membranes as a function of methanol concentration

Swelling ratio as methanol concentration depicted in **Figure 2**.Nafion 117 has the highest methanol swelling ratio Incorporating CaO into PVA matrix resulting in the lower methanol swelling ratio. **Figure** 3 shows the swelling ratio changes in 30% w/w methanol as a function of temperature. The swelling ratio decreased significantly with the CaO content. Therefore, it can be suggested that CaO significantly improved the PVA swelling properties.

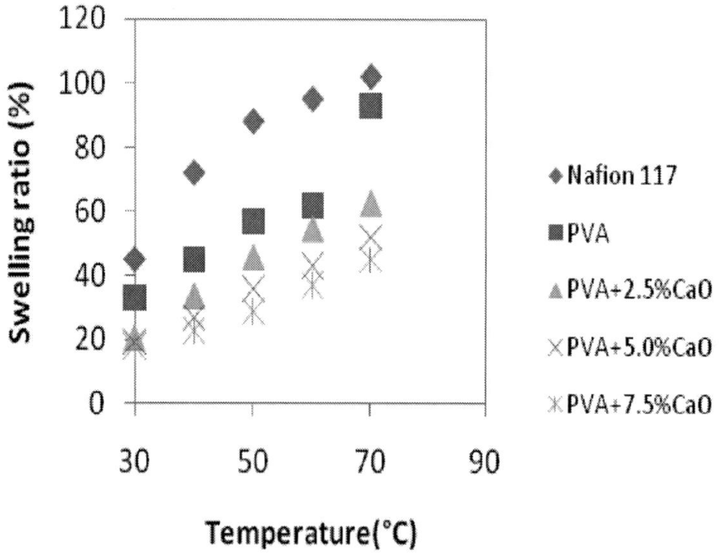

Figure 3: Swelling ratio of membrane in 30%w/w methanol as a function of temperature.

3.3 Methanol permeability test

Concentration of methanol permeated through all membrane samples as function of time in 1.0M methanol depicted in **Figure 4**. Nafion 117 has the highest methanol permeability. Methanol crossover rate decreased with CaO content. The methanol permeability values , P_{CH3OH} were also determined from graph slope and the results shown in **Table 2**. The methanol permeability value for composite membrane is was less than Nafion 117. Therefore, it can be suggested that CaO could enhance methanol selectivity in PVA matrix.

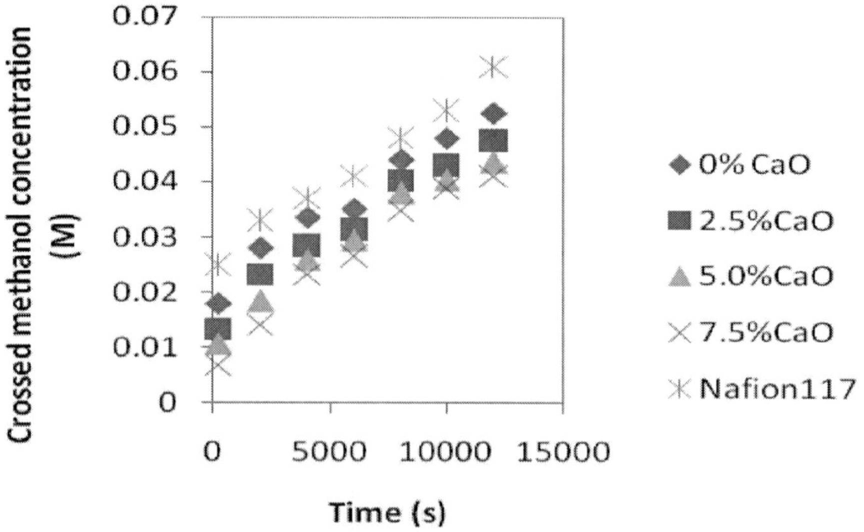

Figure 4 : Crossed methanol concentration in membrane as a function of time.

	σ (Scm^{-1})	P_{CH3OH} (cm^2s^{-1})
Nafion 117	1.14x 10^{-2}	4.94 x 10^{-6}
PVA	1.83 x10^{-3}	3.91 x 10^{-6}
PVA + 2.5% CaO	4.47x10^{-3}	3.33 x10^{-6}
PVA+5.0% CaO	2.52x 10^{-3}	2.27 x 10^{-6}
PVA +7.5% CaO	2.33x10^{-3}	1.813 x 10^{-6}

Table 2: Conductivity (σ) and methanol permeability value (P_{CH3OH}) of membrane

3.4 Conductivity test

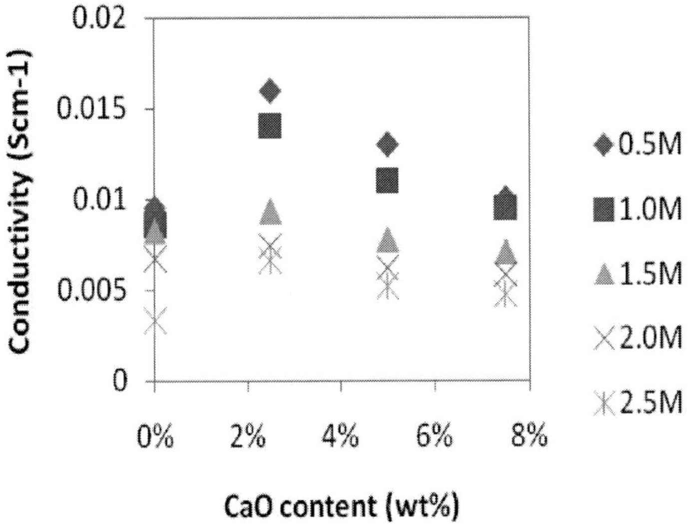

Figure 4: Conductivity value as function of CaO content in various methanol concentrations.

Figure 4 depicts the membrane conductivity as a function of CaO in 5 different methanol concentrations. CaO has significantly improved the PVA matrix conductivity. The conductivity value decreased with the methanol concentration due to the higher methanol crossover rate. Highest conductivity value was obtained in PVA+ 2.5 % CaO. However, at higher CaO content, the ionic transport resistance increased which eventually reduce the ionic conductivity.

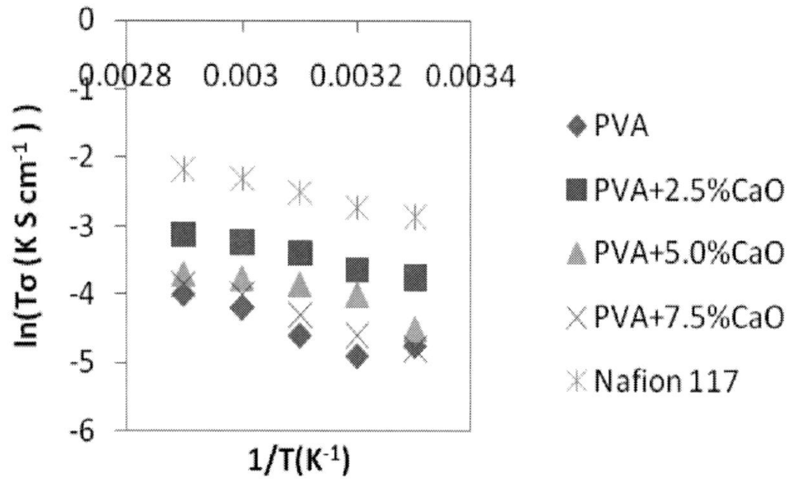

Figure 5 : Conductivity (σ) of membrane as function of temperature

The results for conductivity as a function of temperature are plotted in **Figure 5**. The activation energy for proton conduction was also determined by using the Arrenhenius equation

$$\ln(\sigma T) = \ln \sigma_0 - \frac{Ea}{RT}$$

Where R is the gas constant, is the constant and Ea is the activation energy for proton conduction . All membranes show temperatures dependency of proton conduction. Ea was obtained from the slope and the result shown in **Table** 3. The obtained Ea for Nafion 117 is 10. kJ.mol which is close to other value reported before. The Ea for 2.5 2t% is slightly higher than Nafion. The Ea value increased with the CaO content. All the Ea value are also in the range of 14 -40 kJ mol thus inferred that the proton transport occurred via Grotthes and vehicle mechanism.

Membrane	Ea (kJ/mol)
Nafion 117	5.58
PVA +2.5%CaO	6.63
PVA+5.0% CaO	9.74
PVA+7.5%	10.01
PVA	15.89

Table 3: Activation energy Ea of all membrane

3.5 Single DMFC performance

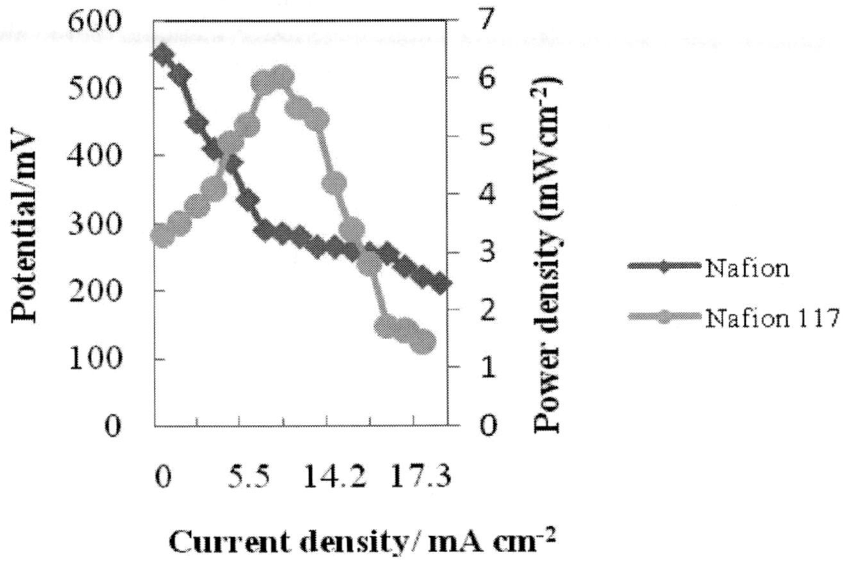

Figure 6: Polarization and power density curves for single DMFC performance for Nafion 117 and PVA+2.5%CaO.

We also determined the possibility of application of PVA-CaO composite membrane by testing it in DMFC and compare the performance with Nafion 117. Only PVA-2.5%CaO test were carried out due to the relatively high proton conductivity value. Polarization and power density curves showed in **Figure 6**. The open circuit voltage (OCV) for PVA-2.5%CaO is higher than Nafion 117 due to the lower methanol crossover rate. The maximum power density attained by PVA-2.5%CaO was 4.29 mWcm-2 which is lower than Nafion 117 (6mWcm-2).

CONCLUSIONS

PVA-CaO composite membrane was fabricated and the effect of CaO particle addition into PVA also was carried out. Fabricated PVA-CaO membrane posses lower methanol permeability compared to Nafion 117. CaO could improve PVA swelling properties and enhance methanol selectivity. At 2.5 wt% CaO showed the moderate conductivity value PVA-2.5%CaO based membrane showed the higher open circuit voltage owing to the lower methanol permeability properties. However, Nafion 117 still exhibited the higher power density and sustained cell performance.

ACKNOWLEDGENT

We are grateful to University Malaysia Sarawak (UNIMAS) for the financial support of this work through Osaka Gas Research Grant. The assistance of EIS measurement from the University Technology Mara also highly appreciated.

REFERENCES

[1] H.Dohle, J.Divisek, and ,R. Jung, (2000), "Process engineering of the direct methanol fuel cell", *Journal of Power Sources,* Volume86, pp469- 477.

[2] A.Heinzel, V.M.Barragan, (1999), "A review of the state-of-the-art methanol crossover in direct methanol fuel cells", *Journal of Power Sources,* Volume 84, pp 70-74.

[3] Y.F.Huang, L.C.Chuang, A.M.Kannan and C.W.Lin, (2009), "Proton-conducting membranes with high selectivity from cross-linked poly(vinyl alcohol) and poly(vinyl pyrrolidone) for direct methanol fuel cell applications",*Journal of Power Sources,*Volume 186, pp 22-28

[4] B.Smitha,S.Sridhar and A.A.Khan,(2005), "Synthesis and Characterization of Poly(vinyl alcohol)-based membranes for direct methanol fuel cell", *Journal of Applied Polymer Science,* Vol 95,pp1154-1163

[5] C. C. Yang, (2008), "Electrochemical performance of an air-breathing direct methanol fuel cell using polyvinyl alcohol/hydroxyapatite composite polymer membrane", *Journal of Power Sources,* Vol 177, pp40–49.

[6]Y. Daiko,T., Kasuga, and M .Nogami., (2006), "Hygroscopic-oxides Nafion hybrid electrolyte for direct methanol fuel cells", *Journal of Membrane Science,* Volume 281, pp 619-625

[7] S.P.Jiang,H.Tang,E.H.Tang,and S.Lu, "Proton Exchange Membrane for fuel cell application", World Intellectual Property Organization Patent No: 136870, 12 November 2009

[8]M. Kouzu, T.Kasuga,M.Tajika, Y.Sugimoto,S.Yamanaka and J.Hidaka., (2008), "Calcium oxide as a solid base catalyst for transesterification of soybean oil and its application to biodiesel production", *Fuel,* Volume *87,* pp2798–2806.

[9]L.T.Nemeth, A.R.Oroskar, K.M.VandenBussche, G.P.Towlerand R.M.Pittman, U.S Patent No: 7,354,461, 8 April 2008.

[10] M.Bello,S.M.Javaid Zaidi and S.U.Rahman, (2008) "Proton and methanol transport behavior of SPEEK/TPA/MCM-41 Composite membranes for fuel cell application", *Journal of Membrane Science,* Volume 322, pp 218-224.

(737a) Photoelectrochemical Properties of Birnessite-Type MnO2 Thin Films for Solar Water Splitting

Thomas F. Jaramillo and Blaise A. Pinaud, Chemical Engineering, Stanford University, Stanford, CA

Photoelectrochemical (PEC) water splitting to produce clean hydrogen fuel is an attractive method for harvesting intermittent solar energy and storing it in the form of chemical bonds. A key photoelectrode material requirement is visible light absorption to maximize the number of photogenerated electrons and holes which can split water ($\Delta E=1.23V$). Manganese oxides (MnOx) are promising photoanodes for PEC cells because they are good oxygen evolution catalysts [1] and the constituent elements are abundant, low cost and non-toxic. While photocurrent generation from MnO2 nanosheets has been demonstrated [2], a myriad of other MnOx phases remain unexplored for PEC. The objective of this work is to explore the photoactivity of birnessite-type MnOx. Thin films of sodium birnessite-type manganese oxide were electrodeposited on fluorine doped tin oxide substrates from a 2mM MnSO4 and 50mM NaClO4 solution [3]. Film morphology was investigated by scanning electron microscopy and crystallinity was studied by x-ray diffraction. Distinct yet broad peaks were identified at 12° and 24° which can be referenced to a crystalline birnessite-type structure with small grain size [3]. X-ray photoelectron spectroscopy was used to determine the oxidation state of manganese in the film [4]. The band gap of the semiconducting thin films was determined by UV-vis absorption spectroscopy; using a Tauc relation the direct band gap of the material was found to be 2.6eV which is larger than the optimal value of about 2eV [5]. Photoelectrochemical measurements were collected using a 1000W Xe lamp and potentiostat. A test cell with 0.1M NaOH was setup in a 3-electrode configuration with the supported MnOx thin film as the working electrode, a Pt mesh counter electrode and a Ag/AgCl (sat. KCl) reference electrode. Anodic photocurrent was detected during chopped illuminated cyclic voltammetry confirming a photoactive, n-type semiconductor. No significant photocurrent degradation was observed after 1 hour of testing at constant potential indicating the film is stable under the cell operating conditions. Preliminary calculations of the band edges based on the flat band potential indicate that the conduction and valence bands of the as-synthesized films straddle only the O2/H2O redox couple; a bias or tandem cell configuration is required to drive the H+/H2 reaction at the cathode. Photoelectron spectroscopy in air was used to confirm the position of the conduction band. Strategies to improve the design and performance of these photocatalysts will also be discussed.

References 1. S. Trasatti, Electrochimica Acta, 29, 1503 (1984). 2. N. Sakai, Y. Ebina, K. Takada and T. Sasaki, J. Phys. Chem B, 109, 9651-9655 (2005). 3. M. Nakayama, T. Kanaya, J. Lee and B. Popov, J. Power Sources, 179, 316-366 (2008). 4. V. Di Castro and G. Polzonetti, J. Electron. Spectrosc. Relat. Phenom., 48, 117-123 (1989). 5. T. Bak, J. Nowotny, M. Rekas and C. Sorrell, Int. J. Hydrogen Energy, 27, 991-1022 (2002).

(737b) Nano Copper Oxide: From Flame Spray Pyrolysis to Photoelectrochemical Hydrogen Generation

Chia-Ying Chiang[1], Sahab Dass[2], Vibha Rani Satsangi[3], Rohit Shrivastav[2], Vidhika Sharma[2], Pushpendra Kumar[2] and Sheryl Ehrman[1], (1)Department of Chemical and Biomolecular Engineering, University of Maryland, College Park, Colleg Park, MD, (2)Department of Chemistry, Dayalbagh Educational Institute, Agra, India, (3)Department of Physics and Computer Science, Dayalbagh Educational Institute, Agra, India

A systematic experimental study of the synthesis of copper oxide nanoparticles and nanoporous films is presented. Powders were made by flame spray pyrolysis method with end applications in the photoeclectrochemical production of hydrogen. Inhibited particle growth was observed for prticles made from 1M copper nitrate (~22%w/w copper nitrate/solution) precursor using a liquid nitrogen gas cooling. The average diameter was 12 nm with a standard deviation of 5 nm. In comparison, particles formed without the additional cooling had an average diameter of 16 nm with a standard deviation of 9 nm. The particle morphology was near spherical and with the fractal dimension (D_f) around 3. With a decrease in the precursor concentration to 0.48% w/w copper nitrate/solution, fractal-type aggregate with fractal dimension (D_f) around 1.9 were obtained, which may be attributed to a transition to diffusion limitation aggregation for the lower precursor concentration. Downstream research into making copper oxide thin film electrodes for photoelectrochemical applications was done by spin coating and doctor blade casting of copper oxide nanoparticles suspension. The behavior of two film preparation process was compared by investigating surface morphology and resulting photocurrent density. In case of films made by spin coating, a uniform surface was observed, however, for films made by doctor blade casting, porous films with diameter of 400nm holes were obtained. The reason of forming different film morphology might be attributed to the evaporation rate of solvent during the drying process. Results for effects of morphology on photocurrent density will be present.

See more of this Session: Nanomaterials for Energy Production and Fuel Cells II
See more of this Group/Topical: Topical 5: Nanomaterials for Energy Applications

(737c) Raman and XPS Investigation of Copper Promoted Transition Metal Doped Ferrites – High Temperature Water Gas Shift Activity

Krishna Reddy Gunugunuri, Chemical and Material Science Engneering, University of Cincinnati, Cincinnati, OH, Punit Boolchand, Department of ECECS, University of Cincinnati, Cincinnati, OH and Panagiotis Smirniotis, Chemical & Materials Engineering, University of Cincinnati, Cincinnati, OH

In our previous studies, we introduced a variety of metal ions (M = Cr, Mn, Co, Ni, Cu, Zn, and Ce) into iron oxide (spinel lattice) and screened for effectiveness for a high-temperature WGS reaction. The idea was to stimulate the ferrite formation via doping with certain foreign cations and to promote the Fe3+ „\ Fe2+ redox couple. In the present study, the role of copper on the shift activity of various transition metal doped catalysts is being investigated. The WGS feed condition were chosen to mimic conditions found in a membrane reactor. For this purpose, various copper doped modified ferrite catalysts were prepared using ammonia assisted coprecipitation method. The prepared catalysts were characterized by using TPR, Raman Mossbauer and X-ray photo electron spectroscopic techniques and evaluated for WGS reaction in the temperature region 400-550 oC and at a stream to CO ratio 3.5 which the conditions used in our membrane reactor. Interestingly, our shift activity findings reveal that copper act as a promoter for the transition metal doped ferrites expect for the Fe/Ce catalyst. While, Cu acts as an inhibitor for the Fe/Ce catalyst. TPR results suggest that addition of Cu to Fe/Ce catalyst promotes the transformation of magnetite to wustite phase to a much lower temperature (starts at 200 oC). This leads to the formation of FeO during the activation of the catalyst which decreases the catalytic activity. Mossbauer spectra of activated and spent catalysts after the WGS reaction reveal that the addition of Cu creates structural distortions in octahedral sites and promotes the WGS activity. In order to find out more about effect of Cu on transition metal doped ferrites in a molecular level, Raman and XPS studies are performed over the activated and used catalysts. These results will be presented in this study.

See more of this Session: Nanomaterials for Energy Production and Fuel Cells II
See more of this Group/Topical: Topical 5: Nanomaterials for Energy Applications

(737d) Placement of Catalyst in Electrospun Silica Nanofibers for the Alkaline Hydrolysis of Biomass

Nathaniel S. Hansen[1], Jeanne E. Panels[1], Ah-Hyung Alissa Park[2] and Yong L. Joo[1], (1)School of Chemical and Biomolecular Engineering, Cornell University, Ithaca, NY, (2)Earth and Environmental Engineering and Chemical Engineering, Columbia University, New York, NY

The use of electrospinning to generate one dimensional materials has garnered significant interest lately due to its inexpensive nature and the diversity of materials fabricated. Recently one dimensional materials have also been investigated to support catalytic nanoparticles to prevent aggregation while maintaining high surface area to volume ratios. In this study inorganic monoaxial nanofibers with nanocrystals dispersed throughout the fiber as well as coaxial nanofibers with nanocrystals selectively placed in the sheath layer have been fabricated through sol-gel synthesis and electrospinning followed by thermal treatment. The fiber morphology and metal domains in silica nanofibers were studied using scanning electron microscopy (SEM), transmission electron microscopy (TEM) and X-ray diffraction (XRD). The resulting silica nanofibers containing metal domains were applied as a catalyst system for the alkaline hydrolysis of biomass to produce hydrogen with minimal carbon monoxide or dioxide contents. Our results demonstrate that a conversion of nearly 100 percent can be achieved, and by placing the catalyst within the sheath layer of the nanofibers, the amount of catalyst required can be significantly reduced (up to four fold).

See more of this Session: Nanomaterials for Energy Production and Fuel Cells II
See more of this Group/Topical: Topical 5: Nanomaterials for Energy Applications

(737e) Composite Intensified Catalytic Membrane for High Temperature Hydrogen Separation

Ahmed M. A. El Naggar[1], Canan Kazak[1], Galip Akay[1], Martin O' Connell[2] and Gunther Kolb[2], (1) Chemical Engineering and Advanced Materials, Newcastle University, Newcastle upon Tyne, United Kingdom, (2)Energy Technology and Catalysis Department, Institut für Mikrotechnik Mainz GmbH, Mainz, Germany

Gasification technology is an important process for the generation of sustainable energy and chemicals technologies from biomass. Separation of hydrogen from syngas generated through gasification is often required in order to circumvent the need to clean syngas which contains tars and heavy metals. Biomass-to-energy/chemicals conversion is more efficient if these processes can be carried out at high temperatures, such as the gasifier exit temperature, i.e., ca. 400 C. Palladium alloys are often used for hydrogen separation. However, due to the nature of syngas, membrane poisoning is unavoidable and the membrane itself should be protected against mechanical and thermal stresses.

In this study, we developed Palladium/Silver membrane modules which are mechanically, thermally and chemically stable. The thickness of the membrane layer is ca. 20-100 µm. Using a model syngas of N2 + CO2 + CO + H2 in the presence of steam and model tar (C30H62), the performance of the porous nickel supported hydrogen selective membranes are evaluated as a function of temperature, pressure and time. Membrane structure was evaluated by Scanning Electron Microscope and XRD.

See more of this Session: Nanomaterials for Energy Production and Fuel Cells II
See more of this Group/Topical: Topical 5: Nanomaterials for Energy Applications

AUTHOR INDEX

AUTHOR INDEX